ELECTRICAL DRAWING I

J. C. Cluley

M

First published 1979 by
THE MACMILLAN PRESS LTD
London and Basingstoke
Associated companies in Delhi Dublin
Hong Kong Johannesburg Lagos Melbourne
New York Singapore and Tokyo

Typeset in 10/12 Times
Printed in Great Britain by
A. Wheaton & Co. Ltd, Exeter

British Library Cataloguing in Publication Data

Cluley, John Charles
    Electrical drawing I.—(Macmillan technician series).
    1.   Electric drafting
    I.   Title
    604'.2'62131042   TK431

    ISBN 0–333–27023–1

Contents

Foreword   vii

Preface   ix

1. Introduction   1

  1.1 Types of Drawing   1
  1.2 Mechanical Drawings   1
  1.3 Symbolic Drawings   2
  1.4 Graphs and Charts   2
  1.5 The User's Needs   2
  1.6 Types of Line   3
  1.7 Sizes of Drawing and Lettering   5
  1.8 Reproduction of Drawings   5

2. Orthographic Drawings   6

  2.1 Types of Projection   6
  2.2 Hidden Lines   8
  2.3 Sectional Views   8
  2.4 Auxiliary Views   10

3. Pictorial Drawings   13

  3.1 Other Types of Projection   13
  3.2 Isometric Projection   14
  3.3 Oblique Projection   15
  3.4 Cutaway Drawings   16
  3.5 Exploded Views   18
  3.6 Sheet Metal Construction   19

4. Dimensioning   21

  4.1 Principles of Dimensioning   21
  4.2 Projection Lines   22
  4.3 Datum Lines   22
  4.4 Functional and Non-functional Dimensions   23
  4.5 Examples of Dimensioning   23
  4.6 Dimensions from Drawings   26

5. Mechanical Symbols and Abbreviations   28

  5.1 Conventional Representation of Detail   28

6. Block Diagrams   32

  6.1 The Function of Block Diagrams   32
  6.2 Block Diagram Layout   33
  6.3 Standard Block Symbols   34
  6.4 Logic Symbols   36
  6.5 British Standard Symbols   37
  6.6 Other Standards   37

7. Circuit Diagrams   40

  7.1 General Principles   40
  7.2 Connection Lines and Connectors   40
  7.3 Symbols for Passive Components   42
  7.4 Symbols for Contacts, Relays and Switches   43
  7.5 Detached Contact Convention   45
  7.6 Symbols for Transistors and Valves   46
  7.7 Symbols for Transducers   49
  7.8 Drawing Circuit Diagrams   49
  7.9 Standard Component Layouts   50
  7.10 Examples of Circuit Diagrams   52
  7.11 Producing Diagrams from Equipment Layouts   53

8. Equipment Design   58

  8.1 Design Assessment   58
  8.2 Design Criteria   58
  8.3 Component Selection   60
  8.4 Types of Resistor   60
  8.5 Types of Capacitor   60
  8.6 Insulating Materials   61
  8.7 Low-voltage Insulation   62

8.8   Materials for High Temperatures and High
      Frequencies   62
8.9   Equipment Testing   62
8.10 Assured Quality Components   63

Appendix A — Standards in Electrical Drawing   65

Appendix B — British and International Standards   67

Bibliography   70

Foreword

This book is written for one of the many technician courses now being run at technical colleges in accordance with the requirements of the **Technician Education Council** (TEC). This Council was established in March 1973 as a result of the recommendation of the Government's Haslegrave Committee on Technical Courses and Examinations, which reported in 1969. TEC's functions were to rationalise existing technician courses, including the City and Guilds of London Institute (C.G.L.I.) Technician courses and the Ordinary and Higher National Certificate courses (O.N.C. and H.N.C.), and provide a system of technical education which satisfied the requirements of 'industry' and 'students' but which could be operated economically and efficiently.

Four qualifications are awarded by TEC, namely the Certificate, Higher Certificate, Diploma and Higher Diploma. The **Certificate** award is comparable with the O.N.C. or with the third year of the C.G.L.I. Technician course, whereas the **Higher Certificate** is comparable with the H.N.C. or the C.G.L.I. Part III Certificate. The **Diploma** is comparable with the O.N.D. in Engineering or Technology, the **Higher Diploma** with the H.N.D. Students study on a part-time or block-release basis for the Certificate and Higher Certificate, whereas the Diploma courses are intended for full-time study. Evening study is possible but not recommended by TEC. The Certificate course consists of fifteen **Units** and is intended to be studied over a period of three years by students, mainly straight from school, who have three or more C.S.E. Grade III passes or equivalent in appropriate subjects such as mathematics, English and science. The Higher Certificate course consists of a further ten Units, for two years of part-time study, the total time allocation being 900 hours of study for the Certificate and 600 hours for the Higher Certificate. The Diploma requires about 2000 hours of study over two years, the Higher Diploma a further 1500 hours of study for a further two years.

Each student is entered on to a **Programme** of study on entry to the course; this programme leads to the award of a Technician Certificate, the title of which reflects the area of engineering or science chosen by the student, such as the Telecommunications Certificate or the Mechanical Engineering Certificate. TEC have created three main **Sectors** of responsibility:

Sector A responsible for General, Electrical and Mechanical Engineering

Sector B responsible for Building, Mining and Construction Engineering

Sector C responsible for the Sciences, Agriculture, Catering, Graphics and Textiles.

Each Sector is divided into Programme committees, which are responsible for the specialist subjects or programmes, such as A1 for General Engineering, A2 for Electronics and Telecommunications Engineering, A3 for Electrical Engineering, etc. Colleges have considerable control over the content of their intended programmes, since they can choose the Units for their programmes to suit the requirements of local industry, college resources or student needs. These Units can be written entirely by the college, thereafter called a college-devised Unit, or can be supplied as a Standard Unit by one of the Programme committees of TEC. **Assessment** of every Unit is carried out by the college and a pass in one Unit depends on the attainment gained by the student in his coursework, laboratory work and an end-of-Unit test. TEC moderate college assessment plans and their validation; external assessment by TEC will be introduced at a later stage.

The three-year Certificate course consists of fifteen Units at three **Levels**: I, II and III, with five Units normally studied per year. A typical programme might be as follows.

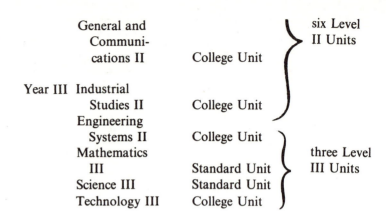

Year I	Mathematics I	Standard Unit	
	Science I	Standard Unit	
	Workshop Processes I	Standard Unit	six Level I Units
	Drawing I	Standard Unit	
	General and Communications I	College Unit	
Year II	Engineering Systems I	College Unit	
	Mathematics II	Standard Unit	
	Science II	Standard Unit	
	Technology II	Standard Unit	

	General and Communications II	College Unit	six Level II Units
Year III	Industrial Studies II	College Unit	
	Engineering Systems II	College Unit	
	Mathematics III	Standard Unit	three Level III Units
	Science III	Standard Unit	
	Technology III	College Unit	

Entry to each Level I or Level II Unit will carry a prerequisite qualification such as C.S.E. Grade III for Level I or O-level for Level II; certain Craft qualifications will allow students to enter Level II direct, one or two Level I Units being studied as 'trailing' Units in the first year. The study of five Units in one college year results in the allocation of about two hours per week per Unit, and since more subjects are often to be studied than for the comparable City and Guilds course, the treatment of many subjects is more general, with greater emphasis on an **understanding** of subject topics rather than their application. Every syllabus to every Unit is far more detailed than the comparable O.N.C. or C.G.L.I. syllabus, presentation in **Learning Objective** form being requested by TEC. For this reason a syllabus, such as that followed by this book, might at first sight seem very long, but analysis of the syllabus will show that 'in-depth' treatment is not necessary — objectives such as '**states** Ohm's law' or '**lists** the different types of telephone receiver' clearly do **not** require an understanding of the derivation of the Ohm's law equation or the operation of several telephone receivers.

This book satisfies the learning objectives for one of the many TEC Standard Units, as adopted by many technical colleges for inclusion into their Technician programmes. The treatment of each topic is carried to the depth suggested by TEC, and in a similar way the **length** of the Unit (sixty hours of study for a full Unit), **prerequisite qualifications, credits for alternative qualifications** and **aims of the Unit** have been taken into account by the author.

Preface

Teachers are frequently reminded that a well-chosen picture is worth a thousand words of text. When describing engineering devices and apparatus this number is surely a gross underestimate. Well-drawn diagrams and scaled drawings convey a wealth of information in a concentrated, precise and easily understood form. Any attempt to replace them by text alone would produce confusion in the reader and almost certainly ambiguity of purpose. For this reason technical drawings are an essential form of communication between engineers, for recording and specifying precisely the details of devices and apparatus.

This book provides the core material required by all electrical and electronic engineers and is particularly intended to meet the requirements of the Electrical Drawing I Syllabus of the Technician Education Council (TEC U75/011).

The preparation and interpretation of engineering drawings and diagrams is a mainly practical activity. Consequently students are urged to supplement their reading of this book with plenty of practice at making drawings and diagrams, and studying carefully any examples which they encounter at work or in other technical studies. This will, I hope, help them to relate the principles given here to the work of professional engineers.

The engineering and drawing Standards mentioned in this book can be obtained from the British Standards Institution, 101 Pentonville Road, London N1 9ND.

J. C. CLULEY

1 Introduction

1.1 TYPES OF DRAWING

Many kinds of drawing are used in electrical engineering in order to record different categories of information. Technicians and technician engineers may have to produce only some of these types of drawing, but they will need to understand all of them, and be able to extract information from them.

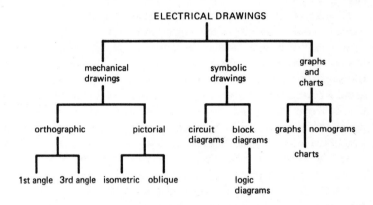

Figure 1.1 Types of electrical drawing

The three main types of drawing, and their subdivisions, are shown in figure 1.1. They are

(1) mechanical drawings
(2) symbolic drawings
(3) charts and graphs

1.2 MECHANICAL DRAWINGS

These are drawn to show the size and shape of an object, or how it is made or put together. They are drawn to scale, that is, there is a fixed relationship between distances on the object and distances on the drawing. The differences between the various kinds of mechanical drawing lie essentially in the number of views shown, the viewpoints used, and the form of projection. Mechanical drawings are described in more detail in chapters 2 and 3.

1.3 SYMBOLIC DRAWINGS

These are drawings such as circuit diagrams which show how components are connected together. They are very different from mechanical drawings because they use symbols to represent the components. The symbols indicate only the *function* of the component and contain no information about its shape, size, or electrical rating.

To take an extreme example, the same basic symbol for an inductor could be used for the single-turn coil which receives the input of a V.H.F. receiver and the massive choke (weighing many tonnes) which may be used to load a 275 kV overhead power transmission circuit. In practice, additional information would usually be added to the basic symbol to show that the V.H.F. coil had a ferrite core, and the mains frequency inductor had a laminated iron core. The symbols used and the way they are assembled into a circuit diagram are shown in chapter 7.

Another form of symbolic drawing in which the symbols denote groups of components is the block diagram. By grouping together a number of separate components into a single functional block, the action of a complicated circuit may be conveyed by showing how the various blocks are connected together.

When tackling an instrument or piece of equipment for the first time it is best to begin by studying the block diagram. This gives an over-all picture of the action of the apparatus. When this has been understood the function of each individual block is much easier to appreciate. For this reason the maintenance and operating hand-books for most electronic instruments and equipment include block diagrams as well as detailed circuit diagrams.

A particular type of block diagram is used to show how digital circuits are connected together. These are called logic diagrams and a particular feature of them is that the symbols used depend only on the logic function performed, and not on the details of the circuit used to construct it. In fact the same logic symbols can be used for non-electrical units such as fluid logic devices. Examples of block and logic diagrams are given in chapter 6.

1.4 GRAPHS AND CHARTS

Graphs express a relationship between two or more quantities and are much easier to use and understand than tables of figures. They are used to specify the characteristics of various kinds of components, particularly active devices. For example, the manufacturer's data sheets on nearly all transistors contain graphs of base current as a function of emitter–base voltage, and collector current as a function of base current.

Charts are similar in appearance but are concerned with more general information. For example, reactance charts show the reactance of capacitors and inductors as a function of frequency. Many charts are produced as aids to design since it is much quicker to read a value from a chart than evaluate several mathematical expressions. Charts of this kind may be used to design filters, attenuators, inductors, etc.

1.5 THE USER'S NEEDS

Another way of classifying drawings is based on the purpose of the drawing and the type of user.

If we consider the life history of a piece of electronic equipment, we find that a variety of drawings will be required during the process of designing, developing, selling, and servicing it. These drawings will go to many different people, each requiring a different set of information, and so a different kind of drawing.

As a rule, the first stage of the project will be the development of a new product and the production of a working model. The design and development engineers will need graphs and charts for the initial design, and in turn will produce graphs showing the model's performance under a range of working conditions. They will also be required to record all the details of the model as a basis for future production. Drawings at this stage are thus mainly recording information.

The next phase is the 'engineering' of the project to make it simple and cheap to make (as far as its basic design allows), and at the same time convenient for the user. For consumer products the final appearance may be very important and may be almost independent of the precise size and shape of the working parts.

In the planning of full-scale production, many detailed drawings are required in order to inform the different sections of the factory of the various non-standard items needed, and usually drawings are sent to outside sub-contractors as well. These are called working drawings, and many are produced so that each supplier or section of the factory will have a drawing for each item needed.

In addition to these, the section which assembles the equipment will need assembly drawings showing how the separate parts fit together, what adjustments are needed, and what clearances between moving parts are required. In the case of spring-loaded electromechanical items such as relays, the assembly drawing may also include the spring tensions at various points.

For sales literature, some outline views of the equipment may be needed, perhaps with performance graphs. When the ultimate user receives the equipment he expects some guide to its operation and how to use it correctly, in the form of an operator's manual. If the user is relatively unskilled, much of the internal detail will be omitted and only the effect of external controls will be described. Simplified drawings may be needed to explain how to perform the essential maintenance tasks such as replacing batteries, making connections to other equipment, cleaning, or lubrication.

Most electrical and electronic apparatus is not repaired by the manufacturer, so to assist the user's technical services (or a specialist repair firm) a maintenance manual is necessary. This should give the complete circuit diagram of the equipment, a description or part number for each component, and the source of any consumable items used. It should also show how to make any adjustments in order to bring the equipment's performance up to the standard of a new product from the factory. Details of any special test circuits or jigs may also be needed.

Unless the equipment is fairly simple, layout drawings showing the position of all the components are usually provided. These are identified by the same designations used in the circuit diagram to assist the service technician to test and replace a component.

The function of the equipment may be illustrated by block diagrams and simplified or signal flow diagrams which show only the signal path and omit minor details such as decoupling, bias circuits and power supplies.

Where the equipment involves some mechanical complexity other drawings may be provided to show how to remove the cover, change certain components, and, for example, how to thread driving belts.

This short survey has included only the main categories of drawing used in electrical engineering, but has shown their great variety. A technician or technician engineer may be called upon to read and understand all of these types, and to prepare a first draft of many of them.

1.6 TYPES OF LINE

In order to distinguish the different features of a drawing, different types of line are used—particularly in mechanical drawing.

The various kinds of line are specified in BS 308: Part 1:1972 and are illustrated in figure 1.2. They have been chosen so that the distinction between them can readily be seen, even when reproduced by various printing and photocopying methods.

To this end only two thicknesses of line are used, the other differences between them being shown by the use of chain-dotted and dashed lines. The thicker continuous lines are used for the main outlines and visible edges of an object. The auxiliary lines such as dimension lines, and the guide lines which lead to them, are shown in thin continuous lines. A line of the same thickness but drawn irregularly is used to indicate the 'broken' or interrupted section of a component of which only a part is drawn.

In order to show the full details of many components it is necessary to indicate edges and outlines which are hidden by other parts, and a special kind of line—the thin dashed line—is reserved for this purpose. Examples of its use are given in section 2.2.

Many components have symmetry about an axis and are produced by turning. An important feature of such parts is the central axis or centre line which is shown by a chain-dotted line. This is thin and consists of alternate short and long dashes. The same type of line is also used to show the end positions of moving parts. Thus when drawing a cycle pump the normal line could be used to draw the pump in the closed position, as it is held in clips on the cycle frame. A second chain-dotted view of the handle, and the

Example	Type of Line	Line Width	When Used
————	continuous thick	mm 0.7	visible outlines and edges
————	continuous thin	0.3	dimension and guide lines hatching, outlines of revolved sections
～～～	continuous irregular thin	0.3	limits of partial views or sections where line is not an axis
- - - -	short dashes thin	0.3	hidden outlines and edges
— - — - —	chain thin	0.3	centre lines, extreme positions of moving parts
▬ — ▬ — ▬	chain, thick at ends and changes of direction, thin elsewhere	0.7 0.3	cutting planes
▬ - ▬ - ▬	chain thick	0.7	surfaces which have to meet special requirements

Figure 1.2   Types of line used in engineering drawing

shaft on which it is mounted, could also be drawn to show how the pump would look if the handle was withdrawn to its full extent.

Another example is the speed-change lever of a tape-recorder. This may have three possible positions, which usually cause an idler wheel to engage with one of the three steps of a stepped pulley. The lever could be drawn in the extreme left-hand position using normal lines, or the extreme right-hand position could be shown by a chain-dotted outline as illustrated in figure 1.3.

The other main use for chain-dotted lines is to indicate cutting planes when drawing sectional views. In order to reduce the number of views of a component it is often possible to combine in

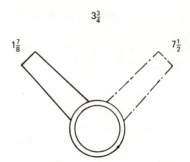

$3\frac{3}{4}$

$1\frac{7}{8}$ $7\frac{1}{2}$

Figure 1.3   Use of chain-dotted lines to show the extreme position of a speed-change lever of a tape-recorder

one drawing sections taken in different planes. This is particularly convenient for turned components which are symmetrical about the central axis. There is thus little point in drawing more than half of a section, since the other half must be identical. The precise position of the section planes is shown by chain-dotted lines in another drawing of the component which gives a side view.

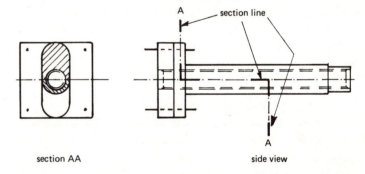

A   section line

section AA   side view

Figure 1.4   Drawing of a moulded coil former—showing the use of section line

An example of sections taken in two planes is shown in figure 1.4 which illustrates a small coil former. This has provision for four pins to connect to a printed circuit card and normally holds a threaded cylinder of ferrite, or 'slug', inside the threaded tubular portion. The coil can be tuned by moving the slug within the former.

In the drawing two half-sections are shown, one through the tubular part and the other through the mounting flange. Note that the chain-dotted line is normally drawn thin, but is thickened at each end and at changes of direction.

1.7 SIZES OF DRAWING AND LETTERING

The standard sizes for mechanical drawings are given in BS 308:1972 and these are also usually adopted for drawings of electronic and electrical equipment. They correspond to the international A sizes, in which each sheet can be divided into two equal sheets of the next smaller size. In order to permit this, the ratio between the long and short sides of each sheet must be nearly $\sqrt{2}:1$. Table 1.1 shows the size of the sheets in both metric and imperial units, and the recommended minimum height for lettering.

Table 1.1

Code	Size in mm	Size in Inches	Minimum Height of Characters Drawing Nos. Titles, etc.	Dimensions and Notes
A0	841 × 1189	33.1 × 46.8	7 mm	3.5 mm
A1	594 × 841	23.4 × 33.1	7 mm	3.5 mm
A2	420 × 594	16.5 × 23.4	7 mm	3.5 mm
A3	297 × 420	11.7 × 16.5	7 mm	3.5 mm
A4	210 × 297	8.27 × 11.7	5 mm	2.5 mm

Note that the table gives only the size of the drawing sheet. To allow some margin when the drawing is reproduced the drawing itself should generally be contained inside a frame. The border between the frame and the edge of the sheet should be at least 15 mm.

Most organisations that prepare drawings in quantity have the drawing sheets printed with a frame and some title blocks and lettering. This gives a standard layout and reduces the amount of lettering which must be added by the draughtsman.

1.8 REPRODUCTION OF DRAWINGS

It is important to note that the lettering sizes given in table 1.1 are recommendations for the final printed version of the drawing. When a photographic process is involved, the size of the drawing may be changed to suit the other material presented with it. Thus in many operating and service manuals drawings may be reduced in size to fit into booklets which are smaller than size A4. The draughtsman who prepares the drawing needs to know how much the drawing size will be changed so that he can make allowances for this in the size of lettering.

Some drawings need to be reproduced to an exact size; for example, those used to make printed circuit boards, which may include plugs to mate with multi-way sockets or the fixed contacts for switches. To allow the size of the reproduction to be checked, special marks are usually drawn at the corners of the drawing which should be a specific distance apart in the final reproduction.

EXERCISE

1.1 Draw an outline view of the following objects using dotted lines where necessary to indicate hidden lines

    (a) a pair of electrical pliers
    (b) a conduit box to mount a lighting switch
    (c) a fuse-holder
    (d) a cassette for a tape-recorder
    (e) a brush for an electric motor.

2 Orthographic Drawings

2.1 TYPES OF PROJECTION

Mechanical drawings are prepared by projecting the outline of the object on to one or more planes. The lines of projection are always parallel in engineering drawings, so that two lines in the object which are parallel and of equal length appear of equal size in the drawing. For example, a set of four parallel edges of a cube will always appear as lines of the same length in an engineering drawing.

The only exception to this occurs in certain architects' drawings that are designed to show how a building will fit into its surroundings. For this purpose it is desirable to give an illusion of depth and distance so that an object of a given size will appear smaller as its distance from the observer increases. This is called a 'perspective' drawing and has the same appearance as an outline tracing from a photograph.

In engineering drawings several views are generally given of the object, projected on planes which are perpendicular to one another. If the projection lines are also perpendicular to the planes, we have orthographic projection. This type of projection generally requires three main views, but if the shape of the subject is fairly simple, and it has some symmetry, two views may be sufficient. The three views are called plan, front elevation and side elevation, and they are projections of the subject from above, from the front and from the side.

A component which can be turned on a lathe, and so has a circular cross-section everywhere, may need only one view to specify its shape completely. However, text added to the drawing or an additional view will be needed if there are other shapes at the ends—such as a slot for a screwdriver or a hexagonal recess for an Allen key.

Unfortunately there are two different conventions for the way in which the projected views are shown, called 1st angle projection and 3rd angle projection. In a simple case it may be easy to decide which projection has been used, but with a complicated shape the decision requires hidden lines to be drawn.

A diagrammatic way of indicating the type of projection is recommended in BS 308. This uses a simple shape—a portion of a cone. Viewed from the side it projects to a trapezium, and viewed

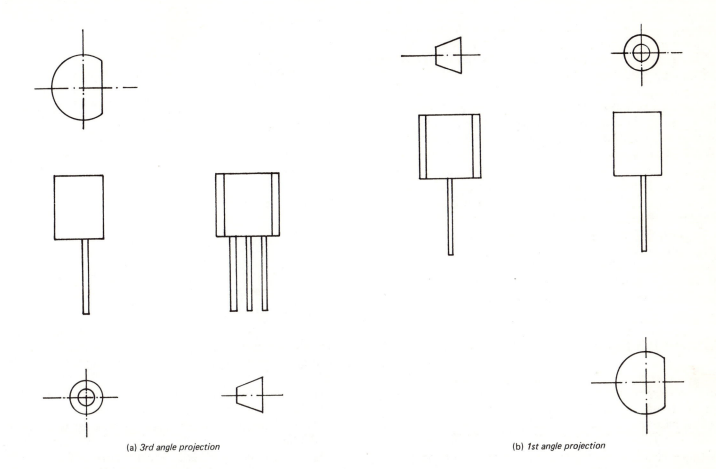

(a) *3rd angle projection*   (b) *1st angle projection*

Figure 2.1   1st and 3rd angle projection

from the top it projects to two concentric circles. If two views of this are drawn alongside the drawing the kind of projection is shown clearly. The view comprising two concentric circles is on the right for 1st angle projection and on the left for 3rd angle projection.

This convention is shown in figures 2.1a and 2.1b, which give three views of a simple shape, drawn with both types of projection. The object shown is the outline of the plastic case used for many low-power transistors. It is deliberately made asymmetrical, so that

there is little chance of its being inserted into a printed circuit board—either by hand or by automatic equipment—the wrong way round. The same shape is used by many manufacturers and is designated as style TO-92.

One way to remember the difference between the two methods of projection is to consider how one view is obtained from the view next to it. If we consider the lower two views in figure 2.1a using 3rd angle projection, we see that the view on the right is obtained by viewing the object shown in the other view also from the right.

Similarly the view on the left is obtained by looking at the object from the left, and the top view is obtained by looking at the object shown below it from the top. The body of the transistor then hides the lead-out wires from view.

The arrangement of the views in the 1st angle projection of figure 2.1b may be obtained simply by imagining the object placed on top of the view shown on the left. If it is rolled to the right through 90° it will appear as in the right upper view, and if it is then rotated above the paper through a further 90° but downwards so pivoting on the ends of the leads, it will appear as shown in the lower view. Expressed in another way, if we start with one view, the view drawn on the right of it shows its appearence from the left, and the view below it shows its appearence from the above.

In general, 1st angle projection is used in Europe and 3rd angle projection in the United States. BS 308 recommends that they should be given equal status, and expresses no preference for a particular projection. It does appear, though, that the use of 3rd angle projection in the United Kingdom is increasing.

2.2 HIDDEN LINES

If the full details of a component are required it is often necessary to draw some of the lines which cannot be seen from outside, or which are on the far side of the component. These are called hidden lines and are shown by using dotted lines. Where there is much information needed about the inside of a component, a sectional view conveys the shape more clearly; this shows what would be seen if the component were cut along some specified plane. The plane along which the imaginary cut is made is usually shown in one of the views.

An example of hidden lines and sectional view is given in figure 2.2. This illustrates a cable lug, which is soldered on to the bared end of an electric cable. The socket end is hollow in order to receive the cable, and the other end is drilled to enable it to be bolted to a busbar or the terminal pillar of a switch, fuse, etc. Although the dotted lines show that the lug is hollow they do not reveal whether this part is cylindrical or square. By drawing an auxiliary view consisting of a section of the socket end we can make

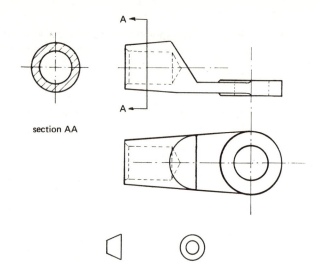

section AA

Figure 2.2 Cable socket

the shape of this portion quite clear.

In practice, the sectional view can be avoided when dimensioning the component since the dimension of the hole into which the cable fits can be marked as a diameter. This implies that the hole is cylindrical.

The drawing illustrates one way of drawing a section, in this case a view separated from the other external views. The section is a view of the component which would be obtained after cutting or sectioning it. The position of the cut and the direction in which the remaining part is viewed is indicated in one of the other views. Here the section line is drawn on the adjacent view, and two arrows labelled A show the direction in which the small ring of material is viewed. The sectional view is then labelled 'section AA' to correspond, and the material which would be exposed by an imaginary cut is cross-hatched at 45°.

2.3 SECTIONAL VIEWS

The section of the cable socket shown in figure 2.2 is at right angles to the axis of the hollow portion, and gives details of the shape of the cross-section. In other situations it may be necessary to show internal details of a component which is symmetrical about its axis.

In this case it is better to take a section through the centre-line of the component. Since the component is symmetrical, both halves of a view projected on to a plane parallel to the axis will be mirror-images of one another, and all information about shape and size is given by either half of the view.

It is then possible to combine two views of the component, one being an external view and one a sectional view, giving, say, a half-section, half-elevation. Figure 2.3 illustrates this. It is a half-sectional view of a coupling device used to attach a standard laboratory microphone to an acoustic cavity for testing artificial ears used in telephone measurements. Since the shape is symmetrical about the axis, only half of the outside view and half of the section are required, and these are combined in the one drawing.

A somewhat more complicated shape, the armature core for a d.c. machine, is shown in figure 2.4. This does not include all of the

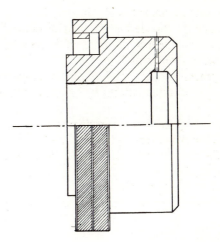

Figure 2.3   Half-section, half-elevation of a microphone coupler

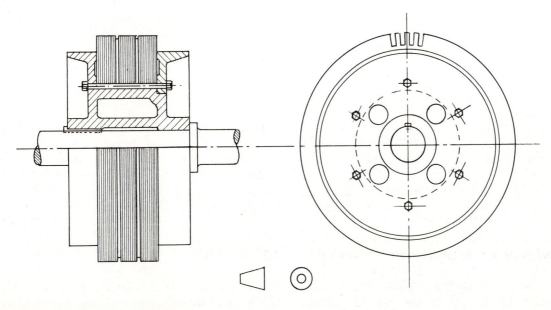

Figure 2.4   Armature core for a d.c. machine

detailed information needed to make the assembly, but is intended to illustrate the various components that are assembled to make the armature core. A working armature requires, in addition to the core shown, insulation for the slots as well as the copper windings which lie in the slots and connect to the commutator mounted on the shaft.

A number of conventions are used in this drawing to reduce the labour involved. First, when drawing the stampings used for the magnetic circuit of the armature, only a few slots are drawn— although the entire circumference of the stamping is slotted. Also, since the stampings or laminations used are so thin (typically around 0.5 mm) it is not possible to draw them to scale, and in the side elevation and section they are shown diagrammatically.

In order to cool the armature, two spacers are included in the stack of laminations. These have radial slots which allow air drawn into the hollow hub to flow outwards to the cylindrical surface and then into the air gap between armature and stator. To avoid confusion the details of these air-ways are omitted.

The end view is in 1st angle projection and shows the view along the shaft of the machine from the left. This includes the head of the key which engages a slot in both hub and shaft and prevents the hub turning on the shaft.

2.4 AUXILIARY VIEWS

Although the three conventional views of an object are generally adequate to convey its shape and size, some additional views may be given to present certain features more clearly. This is often done when a part of the object lies in a plane which is not parallel to one of the three planes of projection. In such a case the part will not be seen directly in any of the three views, and where much detail needs showing and dimensioning, an auxiliary view looking perpendicular to the part makes the drawing clearer.

An example of this is shown in figure 2.5, which gives a plan and side elevation of a bench stand for a panel-mounting meter. The front of the stand is cut away to accommodate the cylindrical body of the meter, and has four holes for the four fixing screws which hold the meter in place.

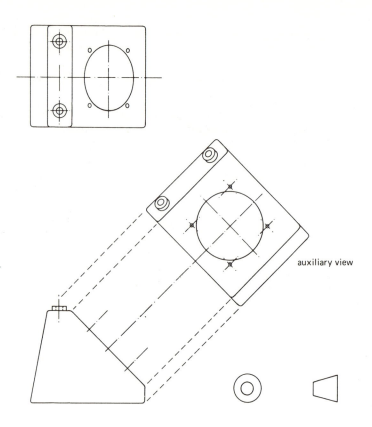

Figure 2.5   Bench stand for a meter

The details and dimensions of the fixing holes and their relationship to the large hole which accommodates the meter movement can be shown more clearly in a view in which the front panel is seen in its correct shape. The lines of projection are then perpendicular to the panel, and the plane of projection is parallel to it. This is shown in the auxiliary view in figure 2.5 which conveys the shape and layout of the panel much more clearly than the oblique view which would be seen in a conventional front elevation.

Another example of an auxiliary view is shown in figure 2.6

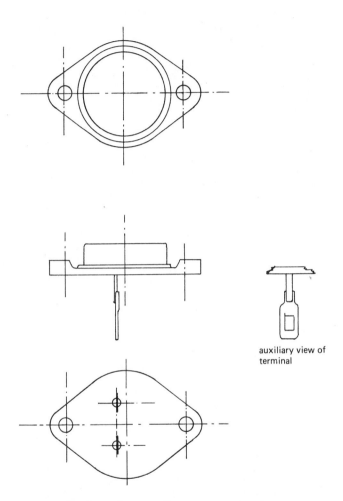

Figure 2.6 Outline of a case for a power transistor

auxiliary view of terminal

the terminals, and contains no redundant information that has already been supplied by the other views.

EXERCISES

2.1 Draw a sectional view of a lamp-holder as used for a domestic 240 V bayonet lamp-cap.

2.2 Draw a plan and side elevation of a dry battery as used in a transistor radio receiver.

2.3 Draw a plan and front elevation of the metal-cased crystal shown in the isometric view of figure 2.7.

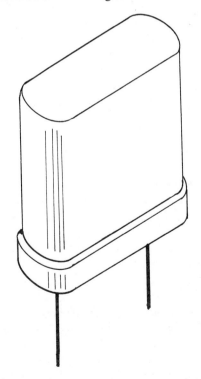

Figure 2.7 Metal can for a crystal

which illustrates a metal-cased power transistor. The three main views provide all the information the user needs about over-all size and shape and the length and position of the leads. An end view looking parallel to the long axis of the transistor would give no further information apart from the detail of the spade terminals. In order to save space and drawing time an auxiliary view is included. This gives only the missing details, namely the shape of

2.4 Draw front and side elevations of the transformer clamp shown in the isometric view of figure 2.8.

2.5 Draw three views of the dual-in-line package shown in figure 2.9 and a sectional view with the section plane shown in the figure.

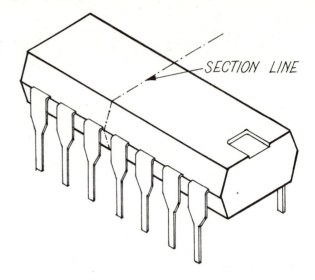

Figure 2.9   Dual-in-line package for an integrated circuit

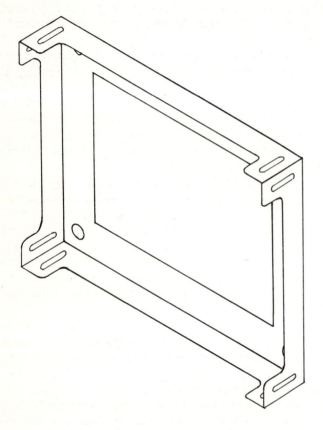

Figure 2.8   Clamp for a small transformer

3 Pictorial Drawings

3.1 OTHER TYPES OF PROJECTION

Although orthographic projection is preferred for drawings which must provide full details of a component for manufacture, it normally requires at least three views for completeness. Where less detail is required, for example in drawings which show how components are assembled or the location of particular features, it is often possible to reduce the number of views which need be drawn.

This is done by using an alternative type of projection in which three different views can be combined in a single drawing. The prospect of replacing three drawings by one is such an obvious simplification that the question naturally arises as to what the disadvantages are. There are two main ones. First, at least two of the three views are projected obliquely and not by lines perpendicular to the major surface, and secondly, the apparent length of a line depends upon its direction.

Inspite of these drawbacks this type of drawing—generally called a pictorial drawing since it gives some impression of depth and perspective—is widely used. It has particular importance for any handbook or catalogue which is intended for the non-technical reader who will be able to follow these drawings easily. This is because they are similar in appearance to photographs which are widely used and understood. In contrast, three conventional views shown in an orthographic projection are not easily understood by the general public.

Examples of pictorial drawings which present information clearly but at the same time simplify the appearance of the subject to some degree are found in most car manuals. These show how to adjust, for example, the carburettor and brakes, and how to change fuses and bulbs, etc.

The particular angle from which an object is seen in a pictorial drawing is generally chosen to reveal its shape and size most effectively. There are, however, two particular types of projection which are used for the great majority of drawings. These are called isometric projection and oblique projection.

3.2 ISOMETRIC PROJECTION

Isometric projection is convenient for shapes which are basically rectangular. If this projection is used to draw a cube for example, the line of sight is inclined equally to all three of the front faces of the cube.

The view of the cube given by this projection is shown in figure 3.1a. The three faces of the cube shown in the drawings have exactly the same size and shape, and the three edges nearest to the observer make angles of 120° with each other. By convention the vertical edge is drawn vertically so that the other two edges are inclined at 30° to a horizontal line on the drawing, as shown.

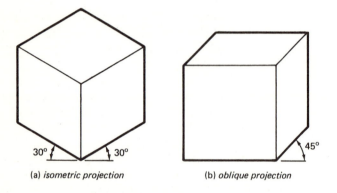

(a) *isometric projection*   (b) *oblique projection*

Figure 3.1   Pictorial views of a cube

The line of sight along which the cube is viewed joins the nearest and farthest corners, and all sides of the cube appear of equal length. Since all edges of the cube are inclined at more than 90° to the line of sight, they appear foreshortened in the drawing. Their apparent length is found by multiplying their actual length by the factor $\sqrt{(2/3)}$ or 0.816.

To help prepare isometric drawings rulers are available which have special scales reduced in this ratio. Thus 1 cm marked on the scale measures 0.816 cm and gives the apparent length of a 1 cm line parallel to any of the edges of the cube.

Also, for preparing free-hand sketches of drawings which can subsequently be traced, isometric graph paper is available. This has a grid of lines parallel to the edges of the cube; because they are evenly spaced the lines form a set of equilateral triangles.

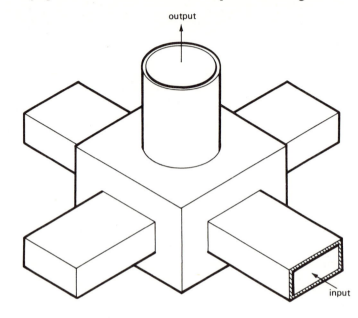

Figure 3.2   Isometric drawing of a turnstile waveguide junction

A component whose shape is almost entirely made from elements which have a rectangular section is shown in figure 3.2. This is a waveguide junction comprising four sections of hollow rectangular tube and one tube of circular section. The power is fed into one of the rectangular tubes, and the other three have closed ends. The circular output connects to a rotating joint (hence the need for a circular section) to energise a rotating radar aerial.

The drawing uses isometric projection and the end of the circular waveguide is a section by a horizontal plane. It thus appears as an ellipse with its major axis (that is, its greatest diameter) drawn horizontally. All of the straight lines in this drawing lie parallel to one or other of the three axes of the drawing and are thus drawn either vertically, or at an angle of 30° to the horizontal.

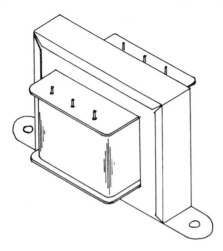

Figure 3.3 Isometric view of a small transformer

A somewhat more complicated shape, but one which is still rectangular, is shown in the isometric drawing in figure 3.3. This illustrates a small transformer with a laminated core, a moulded bobbin containing the windings, and a pressed-steel clamp which holds the laminations and is used to bolt the transformer down to a chassis or mounting plate. The windings are normally concealed by a layer or two of insulation.

3.3 OBLIQUE PROJECTION

Although isometric projection is the most widely used method for pictorial drawings, it has the disadvantage that no face of a rectangular object is viewed directly. Thus no face will be shown with its correct shape or size.

Where there are distinct advantages in showing at least one face correctly an alternative type of projection—called 'oblique' projection—can be used. This favours faces in one plane (generally a vertical plane) which appear full size. Thus both horizontal and vertical lines in the plane are drawn full size, as shown for the front face of the cube in figure 3.1b.

The remaining direction, perpendicular to the front face, is shown by a line at 45° to the horizontal and distances are drawn half full size. Thus in figure 3.1b the six horizontal and vertical edges which can be seen are drawn full size, and the three inclined edges are drawn half size.

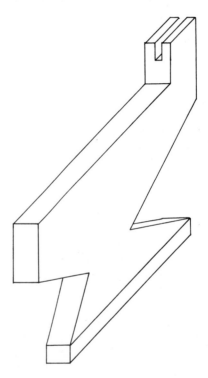

Figure 3.4 Pictorial view of a commutator segment using oblique projection

An example of a component drawn in oblique projection is given in figure 3.4. This is one bar of a commutator for a small d.c. machine. The bar is held between two insulated rings with edges having a V-shaped profile which engage in the V-shaped slots in the bar. The brushes bear upon the top surface, and the armature coil is swaged or brazed into the slot cut in the raised part at the far end of the bar.

This type of projection can always be identified by noting that horizontal and vertical lines in the main face of the object appear as horizontal and vertical lines in the drawing, and the direction at right angles to them appears at 45° to the horizontal in the drawing.

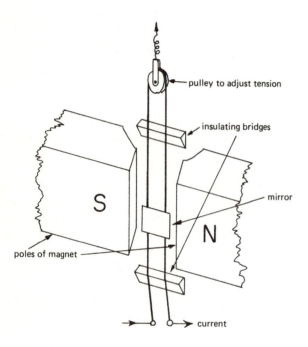

Figure 3.5   Pictorial view of a mirror galvanometer

To illustrate a non-standard type of projection, figure 3.5 has been drawn using a viewpoint from which the observer does not see any surface perpendicular to the line of sight. Vertical lines are drawn vertically, but the major horizontal lines are not drawn at 30° to the horizontal (isometric projection) or at 45° (oblique projection). The main reason for using a non-standard type of projection such as this is to allow a viewpoint which shows most clearly those particular features of the object that are important.

3.4   CUTAWAY DRAWINGS

In some pictorial drawings there is internal detail which in an orthographic drawing would be shown by a sectional view. To avoid the need for a separate drawing the detail can often be shown in a pictorial view, by drawing only parts of the outer components. The drawing then represents an assembly in which some of the outer components have been cut away to reveal the inner detail. A drawing of this kind is usually called a 'cutaway' drawing.

An example is shown in figure 3.6a. This illustrates a small capacitor microphone which is being developed for use in telephones. The metallised diaphragm is shown cutaway to reveal the perforated backplate and the cavity behind it. The body of the microphone is also drawn cutaway over a 90° sector to illustrate the internal construction and the contact spring. Note that the circular shapes of the body and the clamping ring involve circles in a vertical plane parallel to one of the axes of the isometric drawing. Thus they are shown in the drawing as ellipses with their major axes inclined at 60° to the horizontal.

The way in which circles which are parallel to the principal planes of an object appear is shown in figure 3.6b. This illustrates a cube which has a circle drawn on each of the faces nearest to the observer. The circles appear as ellipses with their major axes horizontal (top face) or inclined at 60° to the horizontal (vertical faces). The circular parts of figure 3.6a project to ellipses with the same orientation as the ellipse on the left-hand vertical face of figure 3.6b.

The drawing of figure 3.6a is intended to show the construction of the microphone, and so for clarity some of the finer detail is omitted and certain dimensions are exaggerated—mainly the thickness of the diaphragm and the diameter of the damping holes in the backplate.

A cutaway view of a component having a mainly rectangular shape is shown in figure 3.7. This is an end view of a connector used to connect external wiring to a printed circuit board. The two pins at the top are bent through 90° and mate with pads and holes in the printed circuit cards, the connection being made by soldering. A female connector with spring-loaded sockets mates with the male connector shown. For uniformity the cross-section shown is used

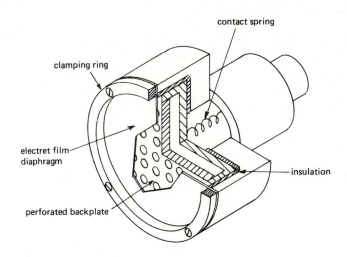

(a) *cutaway pictorial view of an electret capacitor microphone*

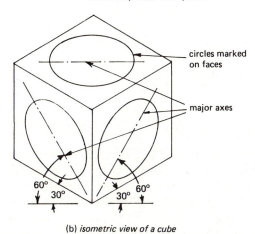

(b) *isometric view of a cube*

Figure 3.6

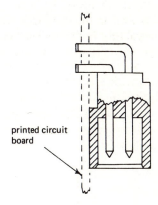

Figure 3.7   Cutaway drawing of a printed circuit connector

allow the mating connector to be inserted more easily.

A cutaway drawing which reveals somewhat more internal detail is shown in figure 3.8. This shows a special quick-acting fuse used to protect high current semiconducting devices from short-circuit faults. The body consists of a ceramic tube which is shown

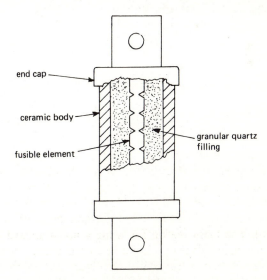

Figure 3.8   Cutaway drawing of a fuse

for several different versions of the connector, each having a different number of pairs of contacts. In order to show the inner details of the connector the outer skirt is shown cutaway, revealing the two rows of pins and the small chamfer on the inner surfaces to cutaway to reveal the granular quartz filling which quenches the arc quickly, and the fusible element which consists of one or more strips of thin silver sheet.

When drawing cutaway views the line along which the cut is made is shown by section lines, but the central detail—for example the contact pins in figure 3.7 and the fusible element in figure 3.8—are shown solid, not sectioned. The boundary line between the solid portion of the component and the cutaway part is drawn as an irregular free-hand line.

3.5 EXPLODED VIEWS

Although cutaway drawings can reveal much internal detail of simple component shapes, they are not very clear when a number of separate components are assembled together. This is largely because of the difficulty of showing that the components are in fact separate items.

The assembly drawing becomes much clearer if the various components are drawn physically separated, but in the same position as when assembled together. This is called an 'exploded' drawing, an example being shown in figure 3.9.

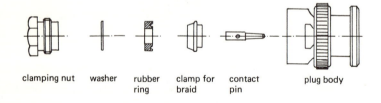

clamping nut   washer   rubber   clamp for   contact   plug body
                          ring   braid   pin

Figure 3.9   Exploded drawing of a plug for a coaxial cable

This depicts the separate components which make up a plug connector for a high-frequency coaxial cable. The cable is threaded through the four components on the left, the outer insulation is then removed, the braid shortened and dressed back against the clamp, and the centre conductor is soldered to the contact pin. The remaining items are then pushed into the body of the plug and clamped together by screwing in the clamping nut.

This has a thread which mates with an internal thread in the plug body. The same components could be drawn as a sectional view but the resultant drawing would be much more confusing.

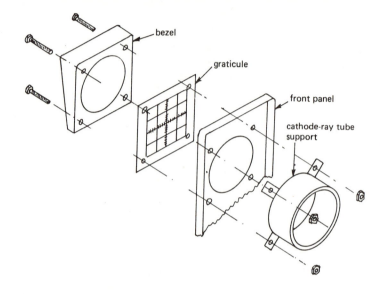

Figure 3.10   Exploded view of a cathode-ray tube

Figure 3.10 shows an exploded view of the supporting ring (usually lined with felt or sponge rubber) for a cathode-ray oscilloscope tube and the way in which it is mounted, together with the transparent measuring graticule and the external bezel on the front panel of the instrument. The items are shown in their entirety except for the front panel which is cutaway because the lower part of it is of no interest in this drawing.

Figure 3.11 gives an exploded view of the components in a clutch for the take-up spool of a tape-recorder. The spindle on which these items are mounted is driven at a constant speed by the main driving motor, and the spool on which the tape is reeled after passing over the recording and replay heads (the take-up spool) is driven by a rubber-tyred wheel which bears against the outer rim of the driving pulley. The main requirement of the drive for the take-up spool is that the tape should be kept under a light, even

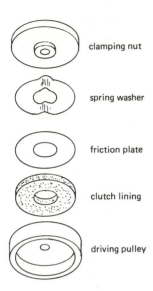

Figure 3.11 Exploded view of a tape-recorder clutch

tension to ensure tight spooling, but the angular speed of the reel must decrease as the diameter of the coil of tape increases. This requires a slipping clutch and in this example the clamping nut is fixed to the spindle by a cotter-pin. The friction pressure is maintained by the spring or 'crinkle' washer, and the lining of cork or a similar friction material transmits a limited torque from the friction plate through the lining to the upper surface of the driving pulley. This drawing is intended as an aid to the maintenance technician who may have to dismantle the clutch to adjust and clean it or to replace the lining; it gives a reminder of the order in which the components are assembled on the shaft.

The exploded view gives a much clearer impression of the shape of the various components than a conventional sectional view and enables them to be identified much more easily.

3.6 SHEET METAL CONSTRUCTION

In experimental or development projects small screening boxes are often required to enclose small electronic assemblies, sometimes with batteries which supply power to them. It is usually convenient to make these as needed from thin sheet metal by cutting and bending. Corners and other parts can be joined by soldering (if using brass, tin-plate, or copper), riveting, or bolts and nuts (if aluminium is used).

In order to decide how to cut the material from the sheet it is helpful to draw an isometric view of the final shape. This can be used to measure the length of various bends and edges in order to produce a cutting plan. A simple example is shown in figure 3.12 in

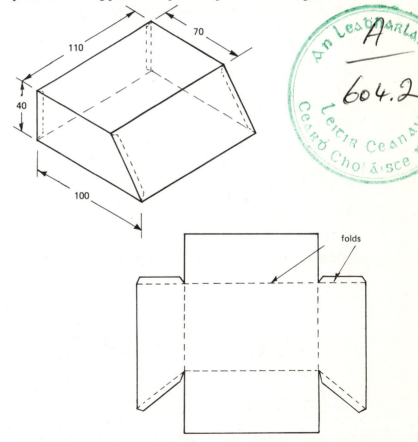

Figure 3.12 Isometric view and development of a sheet metal case

which an inverted tray shape has one side lengthened and set at an angle. This can be used to mount controls and switches and perhaps a small meter. In this example the corners are strengthened by folding small flaps on the sides which are attached to the front and back of the case.

EXERCISES

3.1 Draw an isometric view of the low-voltage, heavy-duty fuse shown in figure 3.13.

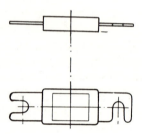

Figure 3.13   Low-voltage heavy-duty fuse

3.2 Draw an isometric view of the fuse-holder shown in figure 3.14.

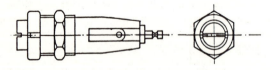

Figure 3.14   Holder for a tubular glass fuse

3.3 Draw a pictorial view of an electric torch to show how the batteries are fitted into it.

3.4 Draw a side view of a 4 mm plug with the cap cut away to show how the wire connection to it is clamped to the end of the pin.

3.5 Draw an exploded view of a 13 A flat-pin plug for domestic use.

3.6 Draw an exploded view of a 13 A flat-pin socket, flush mounting, and the box which is sunk into the plaster of a wall to support it.

4 Dimensioning

4.1 PRINCIPLES OF DIMENSIONING

All mechanical drawings that are used for production must contain dimensions so that the exact shape and size of the component or assembly are fully specified. The amount of information required depends mainly on the shape of the components. A very simple item such as a spacing piece in an assembly may not need a drawing of any kind. Its shape may be that of a rectangular solid cut from a flat bar, and the only dimensions necessary are length, width and thickness. If it is of cylindrical form, length and diameter are needed, and if it is tubular, the internal and external diameters and the length need to be drawn.

Most engineering components have shapes which comprise mainly planes, cylinders or cones and so do not need an excessive number of dimensions. The most difficult shapes to specify are those with more complex curvature; for example, the blade of a steam turbine is curved in three dimensions and requires a large number of co-ordinates to specify its shape—together with some method for developing the shape between the points whose positions have been given. The same difficulty arises in specifying a complex shape such as a car body.

In this book we are concerned generally with simpler shapes which can be specified without a vast number of dimensions. One application which needs special treatment is the profile of a cam which may be used for actuating switches from a rotating shaft. Here the curvature is usually in only one dimension, and a large-scale drawing of the cam profile is usually adequate to define its shape.

Some principles of dimensioning are shown in figures 4.1 and 4.2. Since the change to metric units dimensions are given in millimetres, or metres for large objects. The points or lines between which a dimension is given are joined by a thin continuous line called a dimension line, with arrowheads at least 3 mm long at each end. This line may be broken for the insertion of the dimension, or the dimension may be given alongside it.

The earlier convention was that horizontal dimension lines had dimensions which could be read with the drawing in its normal position, and vertical dimension lines had dimensions which could be read from the right-hand side of the drawing sheet, as shown in

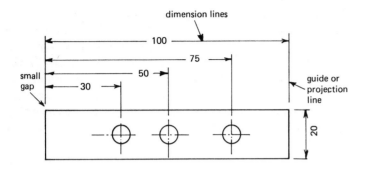

Figure 4.1   Dimensions given from a common datum

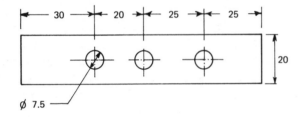

Figure 4.2   Chain dimensioning

figure 4.1. In the interests of clarity, many drawings are now prepared in which all figures can be read from the bottom of the page, as in figure 4.2.

4.2   PROJECTION LINES

To avoid obscuring the drawing it is generally desirable to put all dimension lines as far as possible outside the outline of the object. This requires additional lines called guide lines or projection lines which extend the dimension being indicated to the dimension lines. These projection lines should extend a little beyond the dimension lines but should have a small gap separating them from the corresponding line of the drawing, as indicated in figure 4.1.

Centre-lines which are normally shown as chain-dotted lines should not be used for dimensioning but should be extended by projection lines as shown in both figures 4.1 and 4.2.

The diameters of holes and cylindrical objects should be indicated by the symbol $\varnothing$, with arrows and projection lines as shown in figure 4.1. Earlier drawings often used the abbreviation DIA in place of $\varnothing$.

4.3   DATUM LINES

When a number of positions along a line on the drawing need to be specified, this may be done in two main ways. The first is to choose a datum or reference line on the drawing and dimension all points with reference to it. This method is shown in figure 4.1 for the horizontal distances. The origin of the measurements is sometimes called a datum plane, or surface, since although it appears as a line on the drawing it is generally a plane surface on the object.

The second method of dimensioning is called chain dimensioning. In this method a dimension is given between each point and the adjacent points.

The first procedure which uses a common datum is much better when the accurate location of the features is important. The reason for this can be seen by considering how the object would be machined. For a few items, such as those shown in figure 4.1, the flat strips would be marked out, centre punched and drilled. If we allow a possible error of 0.2 mm in each measurement, the distance of each hole from the left-hand end of the bar would be correct within the same error of 0.2 mm.

If, however, the chain dimensioning arrangement of figure 4.2 were used, the possible error in the distance between the left-hand edge of the bar and the hole farthest from it would be 0.6 mm because each of the three dimensions which together determine the position of the hole could have an error of 0.2 mm. This result of course assumes that the holes are located by marking out the dimensions as given in figure 4.2.

Clearly, in any situation in which the position of the holes is critical, the method of dimensioning shown in figure 4.1 should be used. However, this method occupies more space on the drawing and where the position of the holes is not critical, the chain dimensioning arrangement is more compact. It would, for example, be completely adequate for specifying the position of ventilation holes in a case for electronic equipment where their precise position would not matter within a few millimetres.

4.4 FUNCTIONAL AND NON-FUNCTIONAL DIMENSIONS

Dimensions of many electronic and electrical components can be divided into functional and non-functional categories. The functional dimensions are those which must be controlled to ensure that the components operate as required, are interchangeable with similar items, or will mate correctly with other items to which they are fixed or into which they are plugged. For example, in a standard 13 A flat-pin plug the size, length and separation of the three pins are functional dimensions.

Non-functional dimensions are those which can be altered within limits without affecting the performance of the object or its mating with other items. Thus the size and shape of the insulating body of the 13 A plug are non-functional dimensions, since they could be altered within limits without detriment to the plug. It could be argued that in many examples the length of the pins is not a truly functional dimension since the contact spring in the socket mates with only a portion of the pin. Thus in many designs of socket the mating pin could vary considerably in length without affecting the performance of the socket.

However, it is essential for the safety of the user that no other pin shall be able to make connection with the socket before the earth-pin. It is thus necessary to set fairly close limits on the permitted lengths of the pins. The dimensions given in BS 1363 : 1967 are specified for this reason with a tolerance of 1 mm.

The outlines of thermionic valves, in particular the shape and size of their glass envelopes, can be regarded as non-functional dimensions. The appropriate standard generally specifies only a maximum size for the envelope so that the user can establish the maximum amount of space which need be provided for the valve. Within the maximum size the manufacturer can vary the shape considerably without causing any difficulty for the user.

However, the dimensions of the base—particularly the size and spacing of the pins, and the size, shape and position of the locating spigot—are functional dimensions. They must be controlled within narrow limits to ensure that any valve of given type, made by any manufacturer, can plug correctly into any valve-holder of the appropriate type. The dimensions must be specified accurately since the valve base and the valve-holder are generally made by different manufacturers.

4.5 EXAMPLES OF DIMENSIONING

As an example of dimensioning, the drawing in figure 4.3 gives some outline dimensions of a silicon diode rated at about 250 A. The dimensions are not adequate for manufacturing, but are intended only for the user who wishes to incorporate the diode into a piece of equipment.

The diode is bolted down to a heat sink which dissipates most of the heat generated within the diode. The surface which makes contact with the heat sink is ground flat and is taken as the reference surface. The two electrical connections are made to the base of the diode and to the 16 mm diameter stem at the top of the diode. In order to design the connections the user needs the size and location of the four fixing holes and the diameter and position of the stem on to which is bolted a flexible lead.

Figure 4.4 shows the details of the holes needed to mount four sockets on the rear panel of a rack-mounted assembly. In this case the reference plane is the left-hand edge of the panel and all the horizontal dimensions are measured from this plane. In accordance with the usual convention the over-all dimensions (105 mm and 213 mm) are shown outside all intermediate dimensions. These dimensions are important when ordering the raw material from which the panel is made. Since the height (105 mm) may vary a little because of the difficulty of cutting the material exactly to size, the positions of the two sets of fixing holes 73 mm apart are not specified exactly but are shown as spaced equally from the top and bottom edges of the panel. This is indicated on the drawing by using the = sign for the two dimensions.

Figure 4.5 shows a drawing which specifies the points of contact

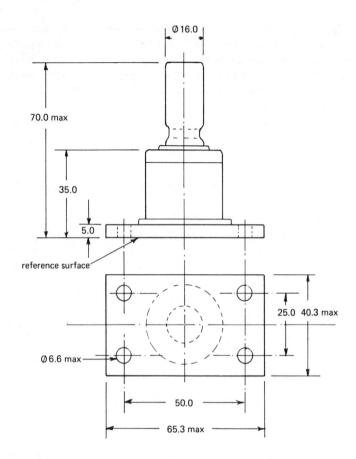

Figure 4.3   Outline drawing of a silicon power diode

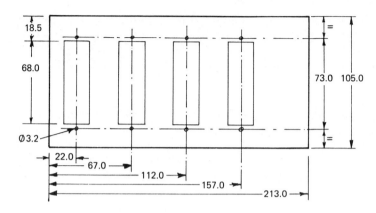

Figure 4.4   Plug mounting details—real panel

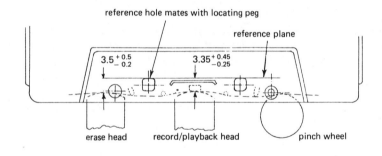

Figure 4.5   Locations of erase and record/playback heads—tape cassette to BS 1568:1973

between the two heads of a cassette recorder and the tape which is held in a removable cassette. This cassette and the mechanical details of the capstan drive, hubdrive, and the location of the cassette within the recorder, are internationally standardised—the relevant document in the United Kingdom being BS 1568 : 1973. The drawing in figure 4.5 forms part of the complete specification, and shows the permissible location of the erase and record/playback heads with respect to a reference plane which joins the highest points of two locating pegs which are fixed to the recorder.

When the cassette is loaded into the recorder two holes in it engage with these locating pegs and the upper surfaces of the pegs are held against the register holes by spring pressure. The same reference plane is then applicable to dimensioning the cassette and the recorder body.

The drawing shows the outside view of a cassette in the loaded position, with the upper surfaces of the heads that project through slots in the front edge of the cassette shown as dotted lines.

The erase and record/playback heads are part of the recorder and to ensure that the cassette plays correctly the locations of the heads are specified within 0.7 mm with respect to the reference plane. In this case the dimension

$$- 3.5 \, ^{+0.5}_{-0.2} -$$

means that nominal value 3.5 mm may be increased to 3.5 + 0.5 = 4 mm or decreased to 3.5 − 0.2 = 3.3 mm without affecting the performance. The tape guides within the cassette will ensure that the tape is kept in contact with the head with this variation of spacing. The standard specification classes the width of the head in the direction of tape motion and the exact head profile as non-functional dimensions.

The maximum size of the head is specified, and provided that the point of contact between the head and the tape is within the limits shown, the precise shape of the head can be altered to suit the manufacturer's convenience. The tolerances could also be indicated by giving maximum and minimum sizes rather than the nominal size with permissible variation, as here. The erase head spacing would then be given

$$\underline{\quad 4.0 \quad} \\ 3.2$$

with the maximum dimension above and the minimum dimension below the dimension line.

Figure 4.6 shows the end plate of a frame for mounting printed circuit cards in a cabinet or a standard equipment rack. The holes

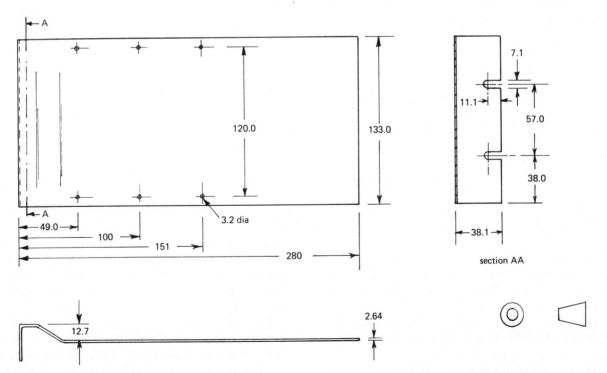

Figure 4.6   End plate for a card frame

locate threaded rods upon which are mounted grooved plastic mouldings and spacers. The mouldings engage with the edges of the boards and hold them in a vertical plane. The main locating surface is the slotted portion which is bent at 90° to the rest of the plate and which is bolted to the front panel by bolts which fit into the slots shown in the sectional view. Accordingly, this surface is used as the reference plane from which all horizontal dimensions are taken.

In the top view the two bends in the left-hand end of the plate are not shown as full lines from top to bottom of the plate since they are radiused curves and not sharp bends. The shape is shown in the plan view and the position of the bends is indicated in the main view by the two pairs of lines drawn partly across the plate.

The details of the two slots are shown in the sectional view drawn on the right. The same detail could be shown by an end elevation drawn on the left of the main view, but if drawn completely this would include dotted lines to represent the main portion of the end plate. These dotted lines would confuse the drawing somewhat and they can be avoided by drawing a sectional view as shown.

Finally, to indicate how angular dimensions may be specified, figure 4.7 shows the view looking at the pins of a B8F base. This can be used either for thermionic valves or for cathode-ray tubes. In order to ensure that the base is inserted correctly into the corresponding holder, a spigot with a key is fixed on the centre-line of the base. The key engages in a slot in the holder and ensures that the base can be plugged into the holder in only one position. The positions of the pins are specified by giving the diameter of the pitch circle upon which they lie (17.45 mm) and then their angular position with reference to the key-way.

A complete drawing suitable for manufacture would also include side views of the pins, showing how the ends are tapered, and of the spigot and key, showing their shape and dimensions.

4.6 DIMENSIONS FROM DRAWINGS

Engineering drawings are normally drawn to scale, and there is thus a direct relationship between the length of a line on the

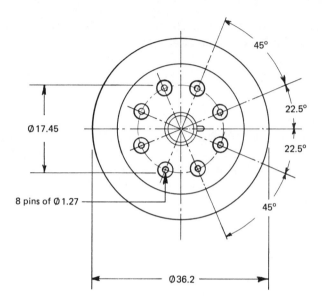

Figure 4.7   B8F base

drawing and the length of the corresponding line on the object. In theory, therefore, any distance not dimensioned on the drawing could be measured with a ruler, multiplied by the appropriate scale factor, and thus give the dimension needed to make the component. Therefore, if a distance measured 35 mm on a half-scale drawing the object would be twice as large as the drawing, and we would expect the dimension on the object to be 2×35 mm $= 70$ mm.

This is, however, a most unwise practice. Normally drawings are prepared on either paper or cloth, and printed on paper, and these materials are much affected by temperature and humidity. Thus the actual size of the drawing may change significantly from day to day, and measurements taken from it are correspondingly unreliable.

Where the drawing is required as part of the production process a more stable material is used. This situation occurs in drawings for printed circuit cards, particularly where these incorporate connectors which must mate with standard multi-way sockets. As an added precaution printed circuit drawings are usually provided with registration marks (at an accurately known distance apart) at the four corners. Such marks give a check on size for full-sized drawings and are invaluable when making negatives for printing printed circuit cards by photographing down original drawings which may be five times full size. An example is shown in figure 7.17.

When reducing drawings photographically some changes in the size of the original drawing can be compensated by varying the reduction ratio. Where this is not possible, for example in making precise maps, metal plates may be used as a drawing material, on account of their very stable size.

Apart from the special cases mentioned above, measurements should *never* be taken from a drawing if any degree of precision is required.

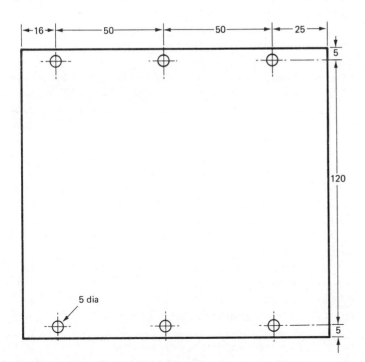

Figure 4.8 Side plate for card frame

EXERCISES

4.1 The drawing of a side plate for a card frame is shown in figure 4.8. Comment on the method of dimensioning the holes and make any changes you think necessary.

4.2 Figure 4.9 shows a plan view of a small chassis which is to be drilled by a numerically controlled machine which uses the top left-hand corner as the datum point for all measurements. How should the dimensions on the drawing be altered to make the setting out as simple as possible?

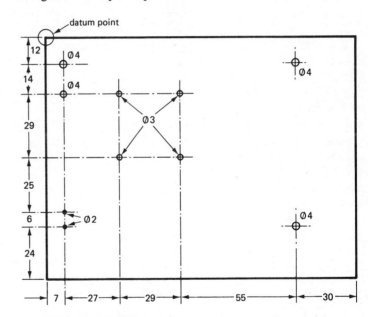

Figure 4.9 Details of chassis

4.3 Add the over-all dimensions, and the dimensions of the position and size of the terminals, to the drawing of the transistor battery drawn for exercise 2.2, chapter 2.

4.4 Draw two orthographic views of a wire-ended electrolytic capacitor, showing the over-all dimensions and the position and size of the leads.

5 Mechanical Symbols and Abbreviations

5.1 CONVENTIONAL REPRESENTATION OF DETAIL

The preparation of mechanical drawings for manufacture or specifications is a labour-intensive activity, and in view of the increasing cost of the skilled manpower needed for this task there is a continuing search for ways of simplifying it and so reducing the time which it involves.

One major method of doing so is by a reduction in the amount of time-consuming detail needed in the drawing, through the use of a form of drawing 'shorthand' to represent items such as screw threads, gearwheels, splined shafts, springs, etc. In order to avoid confusion it is essential that all such simplified details should be standardised so that only a single set of symbols is used and there is no possibility of misunderstanding the features represented. For this reason a set of conventional symbols which represent the standard features is given in BS 308: Part 1. A number of these are shown in figures 5.1 and 5.2.

The first two symbols of figure 5.1 which denote screw threads are particularly interesting; their present form is the result of two stages of simplification. Drawings of fifty or so years ago, when times were less hectic, included almost exact drawings of screw threads, shown in detail that represented each turn of the thread. The only simplification was that the lines representing the tops and the roots of the threads were straight lines whereas they should correctly have been depicted as the projection of spirals.

Thirty years ago, the threads were represented less realistically as a series of long and short lines, with no attempt to draw the serrated outline of the thread. Today the symbol makes no attempt to depict the turns of the thread, and consists essentially of two lines parallel to the axis of the thread, with a further line drawn perpendicular to the axis to indicate where the threaded portion ends.

The present convention has probably simplified the representation of screw threads as far as possible since it is difficult to imagine how any lines could be removed from the drawing in figure 5.1 without losing relevant information.

The symbol for the splined shaft saves detail by depicting only three of of the teeth and giving only the outer and root circles for the remainder. Note that this convention is used for similar

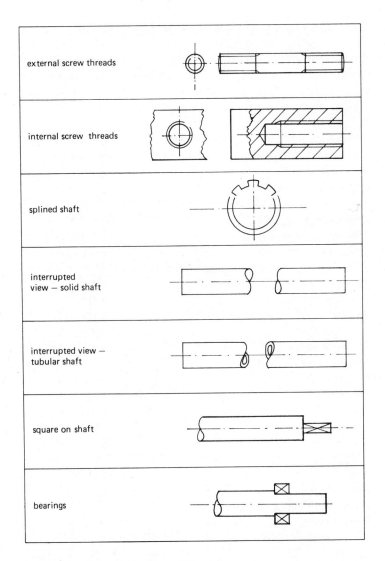

Figure 5.1   Drawing conventions

require a minimum of drawing. In old drawings the broken section of the shaft is usually shaded but the present convention omits

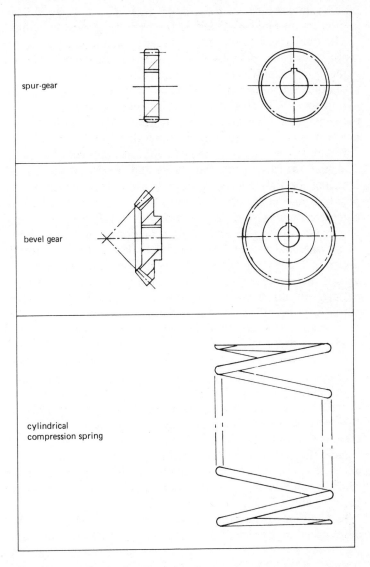

features such as serrated shafts and the profile of armature stampings for electrical machines shown in figure 2.4.

The next two sketches which depict interrupted views of shafts

Figure 5.2   Drawing conventions

this. The convention for a square section on a shaft, although requiring two extra lines on a normal side view of the shaft, saves time because it obviates the need to draw an end elevation or a section to indicate which part of the shaft has a square section.

The final sketch in figure 5.1 shows a simple convention to represent a roller or ball-bearing. These bearings are normally made by a specialist manufacturer and have standard dimensions for both the inner and outer diameters, and the length. Thus the user is not concerned with the internal details of the bearing and the conventional method of representing it shown in figure 5.1 involves only these three outline dimensions.

Figure 5.2 shows further conventions for simplifying drawings. The first two deal with gearwheels and the major feature in reducing the time needed to draw the wheel is that the tooth profile is not shown. If there are many teeth on the wheel, a complete drawing which showed all of them would be very lengthy and tedious to prepare. The only information given is the position on the sectional view of the tip and the root of the teeth. The third sketch in figure 5.2 shows a simplified drawing for a spring. Here, as with the gearwheel, the main outline consists of a repeated pattern—in this case many turns of wire. The drawing shows in detail only the end turns of the spring with an indication of the outer diameter of the remaining turns of the helix.

The usual representations of features shown in figure 5.1 and 5.2 are selected from the larger set given in BS 308: Part 1, which includes many other conventions for mechanical components.

In addition to conventions to simplify drawings, there are many standard abbreviations outlined in BS 308 that are used in the descriptive notes and text incorporated in drawings. A selection of these is given in table 5.1.

Note that generally there is no full stop after the abbreviation unless, as with the abbreviation for number, the abbreviation could be confused with another word. Capital letters are normally used for abbreviations.

In addition to the symbols and abbreviations shown in the table, others may be used for particular purposes. For example BS 499: Part 2 gives a list of standard symbols which denote various types of weld and BS 3939 gives symbols for use in electrical and electronic engineering.

Table 5.1 Abbreviations for Use on Drawings

Term	Abbreviation or Symbol
Across flats	A/F
Assembly	ASSY
Centres	CRS
Centre line	℄ or CL
Chamfered	CHAM
Cheese head	CH HD
Countersunk	CSK
Countersunk head	CSK HD
Cylinder or cylindrical	CYL
Diameter (in a note)	DIA
Diameter (preceding a dimension)	$\varnothing$
Drawing	DRG
External	EXT
Figure	FIG.
Hexagon	HEX
Hexagon head	HEX HD
Insulated or insulation	INSUL
Internal	INT
Left hand	LH
Long	LG
Material	MATL
Maximum	MAX
Minimum	MIN
Number	NO.
Pattern number	PATT NO.
Pitch circle diameter	PCD
Radius (preceding a dimension, capital letter only)	R
Required	REQD
Right hand	RH
Round head	RD HD
Screwed	SCR
Sheet	SH
Sketch	SK

Term	Abbreviation or Symbol
Specification	SPEC
Spherical diameter (preceding a dimension)	SPHERE $\varnothing$
Spherical radius (preceding a dimension)	SPHERE R
Spotface	S'FACE
Square (in a note)	SQ
Square (preceding a dimension)	$\square$
Standard	STD
Undercut	U'CUT
Volume	VOL
Weight	WT
Taper, on diameter or width	$\triangleright$

5.2 Figure 5.4 shows a bracket which supports three spur-gears. The pitch circle diameters are A 40 mm, B 60 mm and C 30 mm. The depth of tooth is 3 mm. Draw the elevation of figure 5.4 showing the three gearwheels using the standard conventions. (The three gearwheels are keyed to the shafts.)

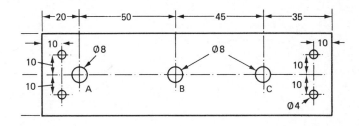

Figure 5.4

EXERCISES

5.1 Figure 5.3 shows a shaft which has a square section 10 mm between flats at one end, an external bearing of over-all dimensions 25 mm outside diameter and 10 mm wide. The other end has a central hole bored and tapped as shown. Draw the view shown, an end view of the square portion, and a section through the threaded portion, using the standard conventions given in figure 5.1.

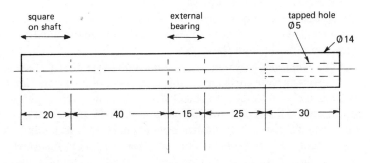

Figure 5.3

6 Block Diagrams

6.1 THE FUNCTION OF BLOCK DIAGRAMS

For manufacturing, commissioning and testing electronic equipment full circuit diagrams are required which show all components and each of their interconnections. However, unless the apparatus is fairly simple and contains no more than, say, 50 to 100 components its mode of operation may be difficult to discover from the circuit diagram alone.

The function of the equipment is much easier to understand if a simplified diagram is available which divides the components into functional blocks and shows only the main signal path through the various blocks. The general operation of the equipment can then be shown, as can the nature of the inputs and outputs of each block. Instead of giving a complete circuit diagram of the apparatus on a single drawing, a separate circuit diagram of each block can be drawn. This makes the circuit easier to follow and eases the problem of handling and storing diagrams.

The contents of any particular block may vary over wide limits depending on the size and complexity of the system concerned. At one extreme, when drawing the block diagram of a radio receiver or an electronic instrument such as an oscilloscope a block may involve only a single transistor amplifier and associated components. In much more complex systems—such as a small computer installation—a single block may denote a magnetic tape store containing many hundreds of components and complex electromechanical parts.

Block diagrams are particularly useful in a descriptive role; for instance, in any publication such as a catalogue, operator's or maintenance handbook in which it is necessary to explain how the equipment works and what changes various signals undergo as they progress through the circuit.

Thus when encountering an electronic unit or system for the first time, an engineer involved in testing, commissioning or repairing it will normally turn first to the block diagram for an over-all picture of the apparatus and the way in which signals are processed by it. Only when this information has been assimilated is it possible to understand fully the function of each separate block, the precise way in which it operates, and the purpose of all the components in it.

Since electronic systems and apparatus are becoming ever more complicated as production techniques are developed, the role of block diagrams in giving a general picture of a system is becoming increasingly important.

6.2 BLOCK DIAGRAM LAYOUT

Block diagrams and circuit diagrams are normally drawn with input signals on the left and outputs on the right so that the main signal path, or paths, flows from left to right. This convention is easy to apply to apparatus which has only a few inputs and outputs but may not be very helpful when laying out the diagram of a complex logic system which has many inputs and outputs. However, even in this instance it is usually possible to arrange the diagram into a number of groups, within each of which the main signal flow is from left to right.

A less complex unit such as a radio receiver can easily be drawn according to this convention by putting the aerial connection on the left of the diagram and the line phones and speaker outputs on the right.

Many block and circuit diagrams, in addition to the main signal handling sections, have auxiliary circuits such as power supplies, control circuits, etc., which are generally located below the main part of the diagram.

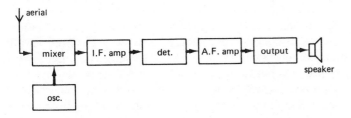

Figure 6.1   Block diagram of a simple superheterodyne receiver

Blocks which have only one or two inputs and outputs can conveniently be drawn with a ratio of breadth to height of 1.5 to 1 (as shown in figure 6.1) but where many inputs or outputs connect

to a single input the clarity and appearance of the diagram may be improved if the shape is altered. For example, the mixer unit in figure 6.3 has been drawn with the height about twice the width. Without this change in shape the six inputs to this block would be crowded together and spoil the appearance of the drawing.

As far as possible the blocks should be spaced evenly over the drawing, avoiding large blank areas. However, where the diagram deals with a system which will be extended later, it is permissible to leave space so that further blocks can be added.

Probably the simplest system which warrants a block diagram is the simple radio receiver shown in figure 6.1. Here the blocks may comprise one or more active devices and the feature of the earlier blocks is that they handle different frequencies. Thus the mixer input is at the variable signal frequency f_1, say, the I.F. amplifier operates at the fixed intermediate frequency f_2, and the oscillator generates a further frequency $f_3 = f_1 + f_2$. The detector extracts the modulating frequency from the I.F. signal and the two remaining stages handle the modulating frequency. This block diagram is useful for testing and fault-finding since it shows clearly the frequencies that should be present at various parts of the circuit.

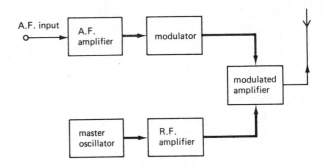

Figure 6.2   Block diagram of an A.M. radio transmitter

A block diagram which contains two parallel channels is shown in figure 6.2. This depicts a small A.M. radio transmitter. The audio amplifier and the master oscillator both operate at low power levels and their outputs need to be amplified in the modulator (audio signal) and the R.F. amplifier (carrier signal) to produce sufficient power to energise the final modulated amplifier stage which feeds the aerial. This stage superimposes or modulates the audio signal upon the carrier wave.

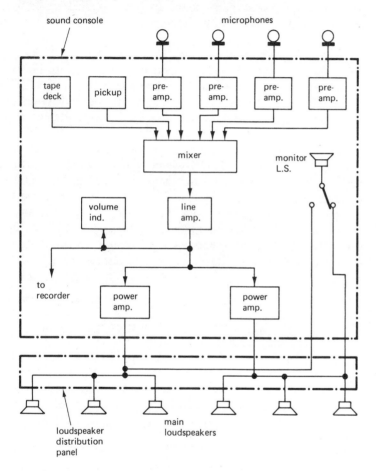

Figure 6.3 Block diagram of a sound reinforcement system

Figure 6.3 shows the block diagram of a small sound reinforcement system which includes considerably more parallel paths than the previous diagrams. The output of the mixer contains a component from each of the signal sources at the top of the diagram, the proportion of each signal being set by a manually operated fader on the mixer control panel. The level of the signal is then increased to feed the two power amplifiers, each of which drives a bank of loudspeakers. In order to make the system as flexible as possible a loudspeaker distribution panel is incorporated which uses patch leads to enable any speaker to be connected to either amplifier. The connections within this panel are shown by enclosing them in a chain-dotted rectangle. Another similar rectangle encloses the units at the top of the diagram which are all mounted in the mixer console. Chain-dotted lines are frequently used in this fashion to show which units in a system are contained in a particular physical enclosure.

6.3 STANDARD BLOCK SYMBOLS

The function of each block in the diagram is usually indicated by a short title printed within the block. This procedure puts little constraint upon the variety of units that can be depicted, but makes the diagram less straightforward to prepare and interpret than it could otherwise be.

Where the blocks in a diagram all depict one of a limited set of units, the interpretation of the diagram may be made easier and quicker by using symbols rather than text to indicate the function of each block.

A particular set of such block symbols is used for telecommunications diagrams, an example of which is given in figure 6.4. This is unusual because it depicts a two-way telephone repeater, which has no specific input or output terminal. Various signals are handled differently in accordance with their frequency bands. Frequencies below 3 kHz are taken to a separate system and are not handled by this repeater, and only signals above this frequency reach the repeater proper through the 3 kHz high-pass filters. Signals originating at the A terminal lie in the range 3–20 kHz and so travel on the upper branch through the 20 kHz low-pass filter, the equaliser, attenuator and amplifier and pass via the second low-pass 20 kHz filter to the right-hand junction and then through the 3 kHz high-pass filter to the line leading to the B terminal. The

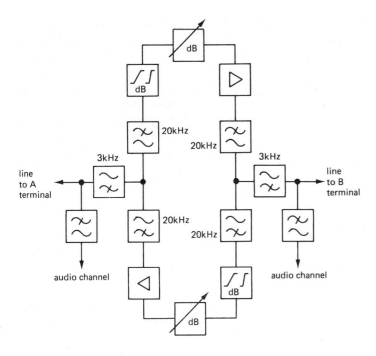

Figure 6.4   Block diagram of a carrier telephone repeater

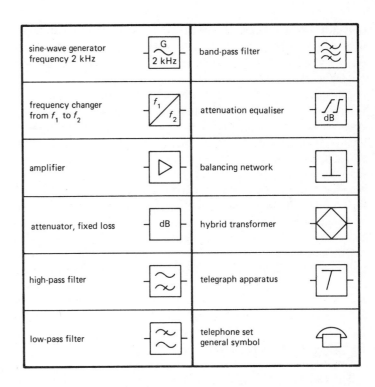

Figure 6.5   Block symbols for telecommunications equipment

two 20 kHz high-pass filters prevent any of this A→B signal from reaching the lower branch.

The B→A signals lie in the frequency band 20–35 kHz and so are prevented from reaching the upper branch by the two 20 kHz low-pass filters. They can, however, travel from right to left through the lower branch as they are readily transmitted through the two 20 kHz high-pass filters. The repeater thus amplifies signals in both directions simultaneously, separating them by the 20 kHz high-pass and low-pass filters. The full set of block symbols used in this kind of diagram is shown in BS 3939 section 22, and the more important ones are given in figure 6.5.

Since there is no need for text to be included in each block to indicate its function the blocks are usually square in shape as shown in figures 6.4 and 6.5. In common with the symbols for electronic components, the property of variation which can be controlled from the front panel of the equipment is shown by drawing an arrow at 45° to the horizontal. Thus, the symbol containing the label dB in figure 6.5 denotes an attenuator having a fixed loss, sometimes called a pad. The same symbol in figure 6.4 with an arrow denotes an attenuator having a loss which can be adjusted by varying a control knob on the front panel of the repeater.

The arrow could also be used with the amplifier to denote an amplifier whose gain could be adjusted by a front panel control. The amplifier symbol occurs so often and in such a variety of circuits that it is often used without the square surrounding the inner triangle.

The two symbols for a telephone set and telegraph apparatus are general symbols, which can be used with other added symbols to denote specific kinds of apparatus, for example a telephone set with a dial, with push button dialling, with a magneto generator, etc., or a telegraph instrument using perforated tape, designed for duplex working, etc.

The equaliser symbols shown apply to attenuation equalisers, which amplify or attenuate different frequencies to a varying extent according to a prescribed specification. They are usually included to counteract the effect of the transmission channel, which in telephony always attenuates the higher frequencies more than the lower ones. The equaliser needs to perform in the inverse manner, that is, to attenuate the lower frequencies more than the higher ones. The effect this equalising has upon the relative phases or the time delay at different frequencies is not specified in this case.

Conversely, in television or telegraph transmission, the relative phase shift between signals at different frequencies is important, as is the time delay down the transmission channel. It is then necessary to include phase or delay equalisers which are indicated by the same symbol but with the letters.

$\emptyset$ or Δt in place of dB

The full title of section 22 of **BS** 3939 is 'Block symbols for telecommunications and general applications'. This recognises current practice in that although the symbols were first standardised for telecommunications equipment only, they are so convenient, particularly as equipment becomes more complex, that they are now widely used in many other branches of electrical engineering.

6.4 LOGIC SYMBOLS

Another special set of symbols which are concerned with a particular branch of electronics are those used to denote logic elements. These are becoming increasingly important as digital techniques are being introduced into nearly all branches of engineering.

The basic symbols were introduced many years ago when the design process was concerned with assembling simple elements together to form a working system (see figures 6.6 and 6.7). The elements were those used in the theory of digital systems, namely, the three logical functions AND, OR, NEGATE, various types of binary storage, and auxiliary functions such as time delay, pulse generators, etc. Comparatively simple symbols can be used to indicate these functions which can be implemented by only a dozen or so components. However, the developments in medium- and

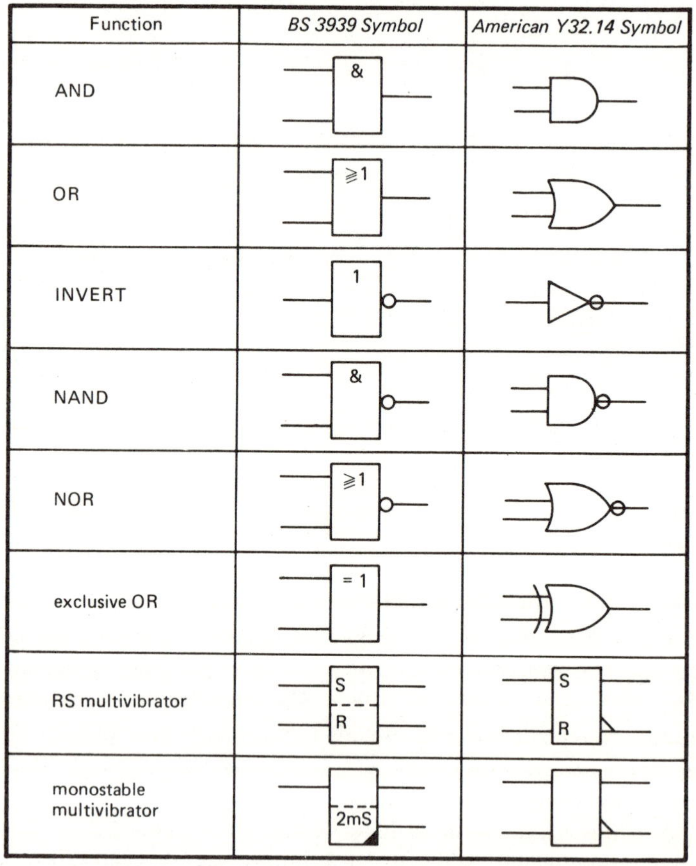

Function	BS 3939 Symbol	American Y32.14 Symbol
AND	&	
OR	≥1	
INVERT	1	
NAND	&	
NOR	≥1	
exclusive OR	= 1	
RS multivibrator	S R	S R
monostable multivibrator	2mS	

Figure 6.6   BS 3939 and A.N.S.I. Y32.14:1973 symbols for logic elements

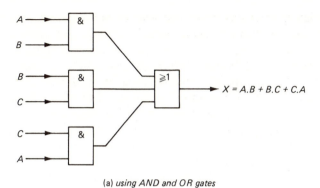

(a) *using AND and OR gates*

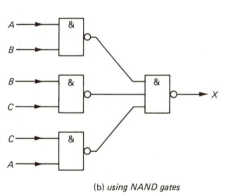

(b) *using NAND gates*

Figure 6.7 Majority voting circuit

large-scale integrated circuits in the last few years have now enabled single packages to be produced which contain many thousands of components. Their functions are consequently very complex and include all of the control and arithmetic operations for a small digital computer.

It is quite impossible to indicate the precise characteristics of these complicated packages by symbols, so that many of the elements in modern logic diagrams must be denoted by rectangular blocks containing text to identify them and only the simple low-level operations use the logic symbols such as those

given in BS 3939, section 21.

The symbols for logic elements have unfortunately not been standardised to the same extent as those for other components and the present situation is a somewhat confusing one. First, we will examine the British Standards for logic symbols.

6.5 BRITISH STANDARD SYMBOLS

It is worth mentioning the various stages in the evolution of the present set of symbols since equipment diagrams that include former symbols are still in use. The first general symbol for a logic element was a circle, which was changed to a semicircle when BS 3939, section 21, was issued in 1969. At this time the move towards a common set of international standards was in motion and in 1971 it was agreed to change the standard gate symbol to a rectangle. The alteration was mentioned in *BSI News* for July 1971 but a revised version of section 21 has yet to be issued. The new rectangular symbol is now used in catalogues and some textbooks—and in later British Standards such as BS 5070: 1976: Drawing practice for engineering diagrams—and agrees with the I.E.C. recommendations.

6.6 OTHER STANDARDS

The most important other standards are those used by American firms, as they are involved in the manufacture and sale of the majority of logic elements now used. The earliest American standard used a fixed rectangular shape for the logic elements, indicating the function by means of text within the blocks. This was considered difficult to understand under adverse conditions by the authorities responsible for servicing equipment in the armed forces. Consequently, another set of distinctive-shape symbols was developed for use in all service equipment and is specified in MIL-STD-806B and MIL-STD-806C. The American semiconductor and computer industry adopted these standards because most of their output went initially for service equipment, and the current American National Standard A.N.S.I. Y32. 14–1973 lists two sets of symbols: the uniform shape set which are substantially the same as those recommended by the I.E.C., and the distinctive-shape set which are with a few exceptions identical to those given in MIL-SPEC-806C.

Thus any electronics engineer who consults catalogues of American manufacturers of logic units, or the circuit diagrams issued by American equipment makers, will need to learn the MIL-SPEC symbols as well as those recommended by the I.E.C. and BS 3939.

Figure 6.6 gives a list of the major logic functions according to both standards. The distinctive-shape set differs essentially in the symbols for the basic logical operations and in subsidiary operations such as time delay. The multivibrator symbols are all based on a rectangular block, but the text and the way in which the leads are labelled differs in the American and BS documents.

Although an international standard set of symbols has now been specified, most American diagrams of logic systems still include the distinctive-shape set. These are somewhat easier to interpret but take considerably longer to draw than the I.E.C. symbols which are all based on a rectangle.

As an example of a simple logic circuit, figure 6.7 shows the diagram of a majority voting circuit. This gives a '1' output if the inputs include two or three '1's, and a '0' output otherwise. It is used wherever extreme reliability is required, and the equipment which provides the input signals is fitted in triplicate. If any one of the three channels fails, the correct signal will still be output from the majority voting unit.

In addition to the variations mentioned above between national standards, some particular manufacturers adopt other variations in the basic gate symbols and in the labelling of signals. For example, a large manufacturer of small computers uses extra labelling on all logic outputs to indicate whether a binary '1' signal corresponds to a high or low voltage.

EXERCISES

6.1 A wide-range electronic a.c. millivoltmeter is divided into the following blocks, given in the order in which the signal travels through the circuit.

(a) 1st attenuator
(b) 1st amplifier
(c) 2nd amplifier
(d) 2nd attenuator
(e) 3rd amplifier
(f) 4th amplifier
(g) meter circuit

There is also a negative feedback path from (c) to (b) and from (g) to (e). Draw a block diagram of the millivoltmeter, showing the main signal path and also the negative feedback paths.

6.2 One method of testing electronic equipment for intermodulation distortion requires two oscillators A and B which each feed into a hybrid network. The output of the network feeds the equipment under test. The output of the equipment under test is taken to a meter M1 (which is connected also to earth) and through a low-pass filter to an amplifier and meter M2. Draw a block diagram of the arrangement.

6.3 Figure 6.8 shows a circuit for adding two binary numbers A and B, drawn using non-standard symbols. Draw the same circuit using the logical symbols of BS 3939.

6.4 Figure 6.9 shows a four-way multiplexor, which feeds the signal A, B, C or D to the output X, depending on the binary signals input to P and Q. This is drawn with American symbols; draw the same circuit using BS 3939 symbols.

6.5 A typical electronically stabilised power unit comprises the following units, starting from the incoming mains supply

(a) a step-down transformer
(b) a rectifier and smoothing circuits
(c) a series regulator transistor

The negative output of (b) is common to the other sections and forms the negative output line. The positive output of (b) is

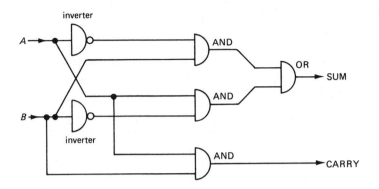

Figure 6.8

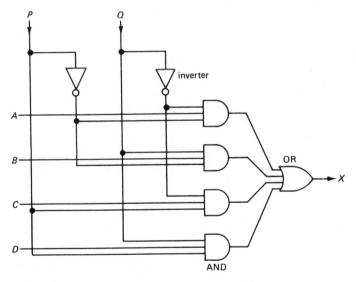

Figure 6.9

of the series transistor (c). The other input to (d) is provided by a voltage reference source (e).

Draw a block diagram of the power unit, including a voltmeter connected across the output terminals.

connected through (c) to the positive output terminal.

Two resistors in series across the output terminals sample the output voltage and their junction is connected to one input terminal of (d) a differential amplifier, whose output feeds the base

7 Circuit Diagrams

7.1 GENERAL PRINCIPLES

Circuit diagrams show the way in which the components in an electrical or electronic system are connected together. When reading or drawing circuit diagrams it is important to remember two points.

(1) The symbol used to represent each component depends only on its function, and has no relation to its shape, size or electrical rating.

(2) The symbols are placed on the drawing to make the diagram as clear and easy to follow as possible. Their position bears no relationship to the layout of the components in the corresponding equipment.

In view of the increase in international trade in electrical apparatus it is important to adhere wherever possible to internationally standard methods of drawing circuit diagrams. This is because much British equipment is exported all over the world, and in particular electronic apparatus may be made in one country, sold in another, and require servicing in a third. Unfortunately there is not complete uniformity in the component symbols used in all countries but the various national standards are steadily being brought into agreement with the international standards recommended by the International Electrotechnic Commission set up for this purpose. The British standard for symbols is BS 3939: Symbols for electronics and electrical diagrams.

7.2 CONNECTION LINES AND CONNECTORS

The connections between components are shown by thin continuous lines, but in some cases the major connections, such as a common earth line, may be drawn in a thicker line to make the drawing clearer. Dotted lines may be used to indicate the position of screening cans or covers and to show mechanical connections. These would be important where several variable capacitors or resistors are connected to a common driving shaft.

Wires that are connected together have a dot at the junction

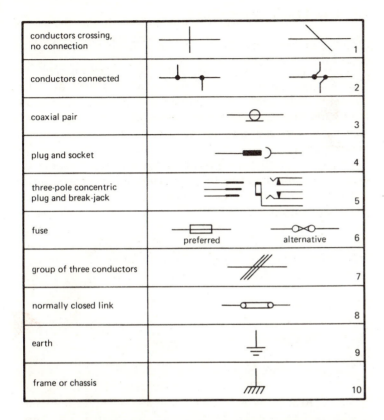

conductors crossing, no connection		1
conductors connected		2
coaxial pair		3
plug and socket		4
three-pole concentric plug and break-jack		5
fuse	preferred alternative	6
group of three conductors		7
normally closed link		8
earth		9
frame or chassis		10

Figure 7.1   Symbols for connections and connectors

point as shown in item 2 of figure 7.1. Note that lines which cross one another as in item 1 should *never* have a dot put at the intersection to denote connection. One of the lines must be altered as shown in item 2 so that there are two three-way junctions and not one four-way junction.

Where a number of conductors run parallel they can be represented by a single line, thus simplifying the diagram; the number of wires concerned is denoted by the number of strokes across the line as shown in item 7. This convention is very useful in diagrams of three-phase power and distribution systems.

The symbol for a coaxial cable, item 3, allows a single line to be used to represent the pair. The outer braided conductor is only drawn at the ends of the cable, apart from its appearance in this symbol.

The plug and socket shown in item 4 connect only one conductor and are usually concentric with one another. In telephone and audio systems there is frequently a need for two or three conductors to be connected simultaneously. The plug has three separate concentric portions: tip, ring, and sleeve, the tip being the shortest of the three lines on the left-hand symbol for a plug (item 5). The plug is inserted into a jack which has a barrel which mates with the sleeve of the plug and two spring-loaded contacts which connect to tip and ring.

The jack shown is a 'break-jack'. This denotes that the spring-loaded contacts are connected to two inner contacts until the plug is inserted. The inner contacts are then isolated and the two outer leads which end in a V-bend connect to the plug.

The two symbols for earth and chassis must be clearly distinguished. The earth symbol must *only* be used for leads or metal parts which are directly connected to a mains earth via a three-wire lead, or are connected to an earth rod (as for example in subscriber's telephone apparatus).

The frame symbol represents the metal framework or the metal chassis upon which the equipment is built, and the use of this symbol is a warning that the metalwork it denotes will *not* necessarily be earthed.

It is thus most important to recognise the difference between the earth and chassis symbols as failure to do so may be dangerous. An example of this hazard occurs in old television receivers where no isolating mains transformer is used and the metal chassis is connected to the neutral wire of the mains supply. This is normally at or near earth potential and so a repair technician may work safely on the receiver when it is switched on. However, if the mains leads have been crossed somewhere the chassis will be connected to the live phase of the mains and will be at a potential of 240 V above earth. Using test equipment with an earthed case to investigate a fault can then be very dangerous.

7.3 SYMBOLS FOR PASSIVE COMPONENTS

The symbols used for passive components such as capacitors, resistors and inductors have to represent a variety of component types. This is done by using a basic symbol to denote the class of component together with an auxiliary symbol to show particular features such as variability, termperature sensitivity, etc.

Component	Symbol	Ref.
basic resistor		R 1
permitted alternative		R 2
variable resistor		R 3
variable resistor — preset	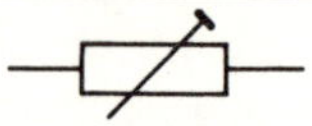	R 4
voltage divider with moving contact	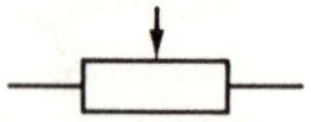	R 5
capacitor		C 6
polarised electrolytic capacitor		C 7
variable capacitor		C 8

Figure 7.2 Symbols for resistors and capacitors

The resistor is a rectangle as shown in item 1 of figure 7.2. This was changed to conform with international standards when section 4 of BS 3939 was issued in 1966, and consequently many textbooks and manuals printed before then will use the earlier zigzag symbol shown in item 2. The general symbol for a variable component is the arrow drawn across the component, as shown in

item 3. This symbol is generally used to indicate that the user or operator of the equipment can alter the resistor value by an easily accessible control.

Where the resistor need only be varied during the manufacturer's production testing—or at rare intervals afterwards for maintenance, or recalibration—it is mounted inside the equipment. It generally requires a screwdriver to adjust it and is called a 'preset' resistor, the symbol for it being shown in item 4.

For many purposes a resistor with a movable tapping point is required; the symbol for this is shown in item 5. This component is used for obtaining a variable voltage or for gain or volume control.

The basic symbol for a capacitor is shown in item 6. Where the capacitor is an electrolytic type, which must always have one plate at a higher potential than the other, the symbol in item 7 is used. The open rectangle denotes the positive plate and the + sign is often omitted. The arrow is used to show a variable component, as in item 8, which would be used for a variable capacitor whose value can be altered by the user or operator of the equipment by using external controls.

Where the capacitor can be altered only by gaining access to the inside of the equipment the preset symbol shown in item 4 would be used.

The basic symbol for an inductor or choke winding is shown in item 1 of figure 7.3. Prior to the publication of BS 3939 the loop symbol of item 2 was used and it is still seen in old textbooks and handbooks. The solid rectangle also shown in item 2 was used mainly in Europe, particularly in Germany, but is nowadays inconvenient because the solid black area does not reproduce well by photocopying.

An inductor with two or more windings is generally used as a transformer and the symbol used is shown in item 4. The polarity of the windings can be shown by the dot convention. This labels the ends of the windings which have the same polarity of induced e.m.f. Thus when the flux in the core of the transformer is changing, the polarities of the marked terminals will both be positive or both negative. If the windings are to be connected so that their e.m.f.s and inductances add together, a marked end of one winding should be connected to the unmarked end of the other winding. Thus if the windings shown in item 4 were the two equal

Component	Symbol	Ref.
basic symbol for winding of inductor or transformer		L 1
alternative, non-preferred	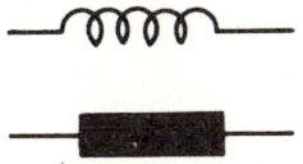	L 2
inductor with ferromagnetic core		L 3
two-winding transformer	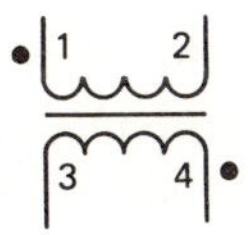	T 4
variable inductor, preset with ferromagnetic core		L 5
inductor with saturable core		L 6
three-phase power transformer star-delta connection		T 7

Figure 7.3   Symbols for inductors and transformers

secondary windings of a mains transformer, terminals 1 and 3 could be connected to two diodes to form a full-wave rectifier, and terminals 2 and 4 connected together to form a centre tap.

Many R.F. inductors contain an adjustable ferrite core, or 'slug', which is threaded and can be screwed in and out of a threaded coil former to vary the inductance. This allows for adjustment of the resonant frequency of the circuit formed by the inductor and some parallel capacitance. The symbol for such a preset inductor is shown in item 5.

For some control purposes it is necessary to use an inductor whose magnetic core is designed to saturate under certain conditions. This usually has at least two windings, one to carry the load current and one carrying a control current. It is generally called a saturable reactor and forms part of a magnetic amplifier. The symbol for a saturable core is shown in item 6.

In large power distribution diagrams all connections involve three-phase conductors and the diagram is unnecessarily complicated if these are all drawn. A special set of symbols are used for these diagrams, given in section 6 of BS 3939, and they usually embody a single line to represent all three phases—the number of conductors being shown by three strokes across the single line as in item 7. This shows the simplified symbol for a transformer comprising two intersecting circles. The connection of the windings is important in this case as it determines the phase shift between the two sides of the transformer. The two symbols inside the circle show that the upper winding is connected in star and the lower one in delta.

7.4   SYMBOLS FOR CONTACTS, RELAYS AND SWITCHES

Many electrical and electronic circuits involve moving contacts and switches, and as with many other symbols, the device is represented by assembling together a number of basic units. The first nine items shown in figure 7.4 are of this kind; the top three may occur on their own, but they are often combined with similar symbols to represent, for example, a three-pole make switch, or a two-pole changeover switch. For the latter the two sections would be similar to item 3 and their moving contacts could be connected together by a dashed line to show that they are mechanically linked

Component	Symbol	Ref.
make contact : general symbol		1
break contact : general symbol		2
changeover contact : break before make		3
relay coil : general symbol	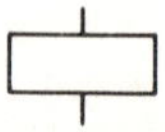	4
coil of slow-releasing relay	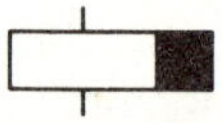	5
coil of slow-operating relay	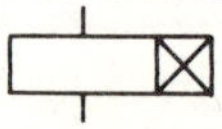	6
relay make contact unit		7
relay break contact unit		8
relay changeover contact unit : break before make		9
three-pole two-position switch	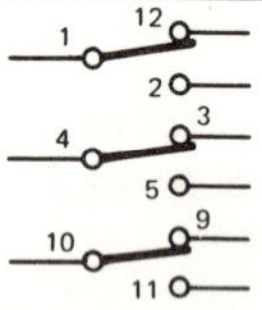	S 10
tag location drawing for above	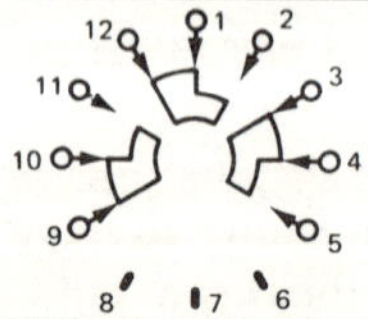	S 11

Figure 7.4   Symbols for contacts, relays and switches

and operate together.

Although electromechanical relays may contain a number of parts, the only items of major interest to the user are the input circuit which consists of an operating coil and the output circuits which consist of sets of contacts which are operated when the coil is energised. Since it is sometimes necessary to know the resistance of the operating coil in order to determine how the circuit works, this value may be written inside the rectangle which represents the coil. By placing a copper sleeve over one end or the other of the iron core inside the coil, the relay can be made slower to operate or slower to release. The symbols representing these features are shown in items 5 and 6.

The output of the relay comprises a set of contacts built up from the three basic units of items 7, 8 and 9. There may be up to four sets of changeover contacts in a typical relay, and more can be fitted in special cases.

A particular type of switch used for low-current circuits in electronic equipment is the multipole wafer switch. It consists of a set of circular switch units operated by a common flattened central rod. This is attached to the control knob and is constrained by a detent or 'click-plate' so that it can only stay in a set of positions 30° apart. A total of twelve contacts can be fitted to the front and the back of each wafer, which can be arranged to have one of the following actions

single pole	11 position
two pole	5 position
three pole	3 position
four pole	2 position

Items 10 and 11 of figure 7.4 show a three-pole two-position switch which requires only nine of the possible twelve tag positions.

In addition to the basic switch types listed above, more complex units can be assembled by shaping the moving contacts. A feature often required when switching R.F. coils is provision for shorting to earth all coils not selected. This is usually provided by an additional rotor which gives the earthing connection to all tags except that selected by the main rotor.

When several wafers are fitted to one switch assembly the switch sections are numbered from the front panel, each wafer having a front section and a back section. The order is thus

SA1F first wafer, front section
SA1B first wafer, back section
SA2F second wafer, front section
SA2B second wafer, back section, etc.

All of these would form part of the complete switch assembly identified by the component reference SA.

7.5 DETACHED CONTACT CONVENTION

In general all the components whose symbols have been given above are drawn together as a single item; for example, the various windings of a transformer are usually drawn next to one another.

Switches, and particularly relays, form an exception to this rule. In simple electronic systems which include only one or two relays and few contacts it is possible to draw the contact units next to the relay coils. However, in more complicated circuits this would create a very complicated diagram which would be both difficult to draw and to follow.

In these circumstances the detached contact convention is used, in which the operating coil of a relay can be separated from the various contact sets which it actuates. Each item is placed in the most convenient position to make the circuit diagram simplest, and easiest to follow. The various contact sets and the operating coil are labelled similarly so that they can readily be recognised as part of the same relay unit.

Thus the relay coil could be labelled RLB/3. This denotes a relay which has three contact units which would be labelled RLB1, RLB2 and RLB3.

The detached contact convention is invariably used in diagrams of telephone switching systems in which most of the components are relays or other electromechanical devices. Without its use such diagrams would become such a mass of lines as to be almost incomprehensible.

The same detached convention is used for circuits in which a number of contacts are operated by a single knob or lever. An

example of this occurs in domestic tape-recorders where a single head, and usually a single amplifier, is used both for playback and recording. Also, the output stage may be used either to drive a loudspeaker or the bias/erase oscillator. There may in all be more than a dozen changeover units operated when the 'record' lever is pressed. The detached contact convention enables the various units to be placed next to the parts of the circuit they control, so making the diagram much easier to read. An example of this is shown in figure 7.5.

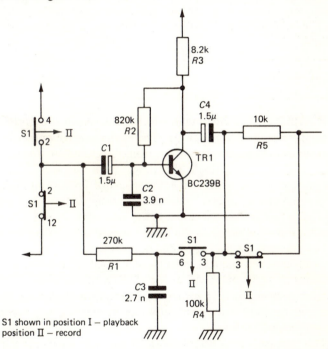

Figure 7.5 Part of a circuit diagram of a tape-recorder showing the use of the detached contact convention

Here, all the switches are operated by the 'record' key which actuates the switch S1. The contacts are numbered in the diagram to correspond with the numbering on the switch assembly. The switch is shown in the playback position I, with contacts 2–4 and

3–6 open and 2–12 and 1–3 closed. When the record key is pressed, the switch is moved to position II with contacts 2–4 and 3–6 closed, and 2–12 and 1–3 open.

The complete diagram contains sixteen pairs of contacts and it would become almost illegible if all of them were placed alongside one another rather than with the part of the circuit in which they are connected.

In order to help to trace signals when fault-finding a summary of the switch action is often provided in the margin of the drawing. This contains a list of the contacts which are connected together in each switch position. For figure 7.5 this would be drawn up as follows

Position I—playback	2–12	1–3
Position II—record	2–4	3–6

A further type of drawing in which the detached convention is convenient is the logic drawing. The equipment they describe is now almost invariably assembled from integrated circuits which may contain several logic functions. Each function is quite independent of the others in the same package, apart from sharing a common power supply. It is thus convenient to draw each logic element separate from the others in the same package and indicate that they are contained in the same package by labelling them.

For example, the package could be identified as IC3 on a layout drawing and the separate elements could be labelled on the circuit diagram as IC3a, IC3b, IC3c, etc. Unless the symbols for the different elements in the package are separated on the drawing and placed in the most convenient positions the drawing may become very difficult to follow.

This process is only useful for small packages which contain a number of electrically separate units. For M.S.I. packages where there is usually a complex circuit with a number of interconnections—for example a four-stage binary or decade counter—the package is drawn as a single rectangle with a single component reference.

The function of the package and its type number may be given alongside the package or a separate table in the margin of the drawing.

7.6 SYMBOLS FOR TRANSISTORS AND VALVES

In section 7.4 we examined how the symbol for a complete relay could be assembled from the various symbols for its individual parts.

This procedure is carried further when assembling the symbols for an active (amplifying) device such as a transistor or a thermionic valve. BS 3939 section 20 gives symbols for the various electrodes of a transistor and some examples of the way in which they can be combined. Although this always shows a circle surrounding the symbol (to denote the envelope) the circle is sometimes omitted as it gives no information where plastic-encapsulated transistors are used. Almost invariably metal-canned transistors have the can connected to one electrode, and this can be shown by a dot at the joint between the line representing the external lead and the circle representing the envelope, as in item 4 of figure 7.7. This is important when fault-finding as the metal can is a very convenient place for measuring signals by a voltmeter or oscilloscope probe.

Figure 7.6 shows the more important symbols for the elements used to construct diode and transistor symbols, and figure 7.7 shows these assembled into typical device symbols.

Figure 7.7, item 1, shows the basic symbol for a p–n diode. Special diodes can be constructed to break down suddenly at a particular reverse voltage, and used for voltage reference elements or for over-voltage protection. The symbol for these zener diodes is shown in item 2. The familiar symbols for junction transistors are shown in items 3 and 4; the dot in item 4 shows that the metal envelope is connected to the collector electrode. Item 4 is marked cbe to indicate collector, base and emitter but this is not usually necessary since the shape of the electrodes indicates their function.

A cathode-controlled thyristor is shown in item 5. This is a four-layer device which has the characteristics of a switch—either fully conducting or open circuit—and the symbol for it is thus quite different from that for a transistor.

In addition to the earlier p–n–p and n–p–n (bipolar) transistors, field effect transistors are now used for a great range of applications, particularly in digital equipment. These control the transistor current by varying an electric field rather than the

Part	Symbol	
envelope		 1
ohmic connection to semiconductor		 2
p-n junction		 3
junction gate of FET — n channel		 4
junction gate of FET — p channel		 5
insulated gate of FET		6
p-emitter on n-type region		 7
n-emitter on p-type region		 8

Figure 7.6   Symbols for the parts of semiconductor devices

current injected into a semiconductor junction.

Since the operating mechanism is quite different from that of bipolar transistors, so is the symbol which denotes them. Many different types can be produced. The gate may be separated from the channel by a reverse-biased junction or a layer of insulation, so producing a junction gate or an insulated gate. The channel may conduct well with zero gate bias and require reverse bias to turn off current; this is a depletion-mode transistor. Alternatively the channel may not conduct at zero bias and instead require forward bias to allow current to flow. This gives an enhancement-mode transistor. Finally, the substrate can be connected to one end of the channel or brought out separately, and either one or two gates can be deposited on the channel. Three of the many types of FET are shown in figure 7.7, items 7, 8 and 9.

Component	Symbol	Ref.
p-n diode		D_1
zener diode		D_2
p-n-p transistor		TR_3
n-p-n transistor, can connected to collector		TR_4
cathode-controlled thyristor	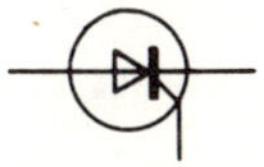	CSR_5
unijunction transistor with p-type base		TR_6
junction-gate FET with n-type channel		TR_7
depletion-type n-channel insulated-gate FET		TR_8
enhancement-type n-channel insulated-gate FET		TR_9

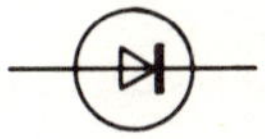

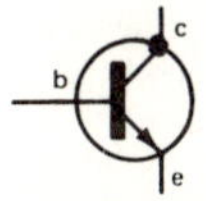

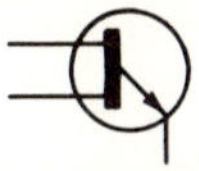

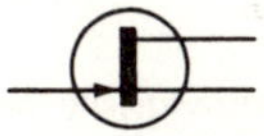

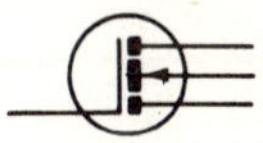

Figure 7.7   Symbols for complete semiconductor devices

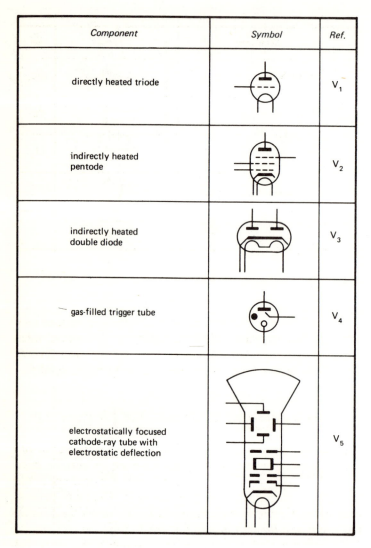

Component	Symbol	Ref.
directly heated triode		V_1
indirectly heated pentode		V_2
indirectly heated double diode		V_3
gas-filled trigger tube		V_4
electrostatically focused cathode-ray tube with electrostatic deflection		V_5

Figure 7.8   Symbols for thermionic valves cold-cathode and cathode-ray tubes

A transistor which has a trigger characteristic is shown in item 6: the unijunction transistor. In this the channel is normally non-

conducting, but as the base voltage is raised it suddenly becomes of very low resistance when the base voltage reaches a critical value.

The symbols for thermionic valves are assembled from a set of basic symbols which represent the various electrodes. All valves are assumed to be contained in an evacuated envelope. Where a valve is operated in a low-pressure gas or vapour its characteristics are substantially altered and this is denoted by drawing a black dot inside the envelope.

Figure 7.8, item 1, shows a directly heated triode valve in which the lowest electrode acts both as heater and cathode. Items 2 and 3 show indirectly heated cathodes in which the cathode which emits electrons is insulated electrically from the heater electrode.

Cold-cathode valves can be operated in a low-pressure gas but have only switching characteristics. A triode of this kind is shown in item 4. This is normally operated so that it is non-conducting until a pulse on the trigger electrode causes an anode–trigger

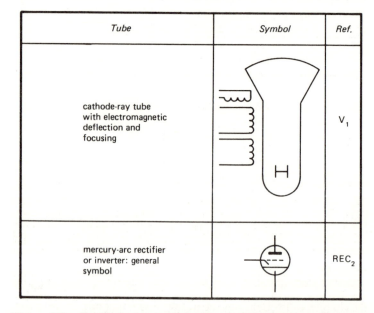

Tube	Symbol	Ref.
cathode-ray tube with electromagnetic deflection and focusing		V_1
mercury-arc rectifier or inverter: general symbol		REC_2

Figure 7.9   Symbols for a cathode-ray tube and mercury-arc rectifier

discharge which in turn causes anode–cathode conduction.

Cathode-ray tubes are of two main types: electrostatic tubes used for osilloscopes and magnetically deflected tubes used for television displays and computer terminals. Item 5 in figure 7.8 shows the symbol for an electrostatic tube which, in addition to a grid to control the beam intensity, has three anodes to provide beam focusing and four plates to provides deflection.

Figure 7.9 shows the symbol for a magnetically deflected tube which also has magnetic focusing. This symbol is simpler since the tube contains only the electron gun (heater, cathode and grid) and final anode. The focusing and deflecting coils are outside the tube envelope.

The second item in figure 7.9 is the symbol for a mercury-arc rectifier in which the cathode is an extremely hot spot on the surface of a pool of mercury. There are often six separate anodes each with a control grid, and there is usually a starting electrode, shown in the symbol, which dips into the pool and then springs back, so drawing out an arc and starting the tube. The same device can also form part of an inverter, which converts d.c. to a.c.

7.7 SYMBOLS FOR TRANSDUCERS

Electronic methods of amplifying and processing signals are very widely used in all forms of engineering measurement on account of their convenience and flexibility. An essential part of this process is the conversion of other forms of energy into electrical energy, or the reverse process.

The device which performs this function is called a transducer, and figure 7.10 shows the symbols for some typical devices. Of these, the microphone and vibration pickup convert acoustic or mechanical energy into electrical signals; the loudspeaker and earphone convert electrical energy into sound waves; and a magnetic tape head may have either electrical input or output. It converts electrical signals to a magnetic pattern if recording and the reverse if replaying. A vibrator acts like a large loudspeaker and turns electrical energy into mechanical vibrations.

The symbols shown in the previous sections are only a small sample of the very large number given in BS 3939 but are typical of

Component	Symbol	Ref.
microphone		MIC_1
earphone (receiver)	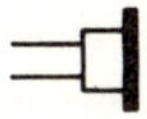	TL_2
loudspeaker	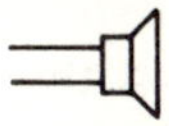	LS_3
vibration pickup or vibrator		4
head for magnetic tape : general symbol		5

Figure 7.10   Symbols for transducers

those most commonly used. The symbols used in specialised areas of electrical engineering, such as those for microwave tubes, television camera tubes, etc., have been omitted and BS 3939 should be consulted for the complete set of standard symbols.

7.8 DRAWING CIRCUIT DIAGRAMS

Although some circuit diagrams are prepared as a record of a particular product, the majority are intended as working drawings. These will be used for manufacturing the equipment they describe, commissioning, or maintaining it. The main purpose, then, of the drawing is to communicate to the reader the details of the circuit and the way in which it functions.

Consequently, when preparing the diagram a very important factor is the clarity with which it shows the circuit's function. This must be given priority over a symmetrical appearance or a uniform spacing of symbols through the drawing.

The following points help to make the drawing easy to read.

(1) The main signal path should run from left to right. In

most electronic equipment one or more inputs and outputs can be identified and this layout is easy to implement. It becomes more difficult with some electrical drawings; and power-station diagrams, for example, are often drawn with the main busbars near the top of the drawing and the various feeds in and out drawn at right angles, that is, vertically up and down the diagram.

(2) D.C. power supply lines should be drawn horizontal, the most positive line being at the top of the page and the rest in order of their potential, the most negative being at the bottom.

Where only one supply line is used this is sometimes drawn above the earth line even when negative to it. For a small diagram involving linear signals this procedure is acceptable, although not always easy to follow.

However, with pulse or digital circuits it becomes extremely confusing when fault-finding with an oscilloscope. This is because oscilloscopes are always built so that a positive-going signal moves the beam upwards. Thus in figure 7.11, with no input signal the transistor base is at the same potential (earth) as the emitter and so the collector current is very small and the collector rises to the same potential as the supply line, $+10$ V.

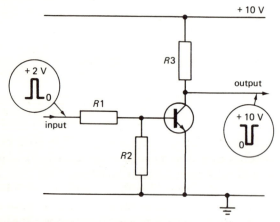

Figure 7.11   Transistor pulse amplifier

A positive-going signal, if the resistors are properly chosen, will drive the base into conduction and saturate the transistor, causing

it to conduct fully so that the collector voltage falls to near earth potential. The two waveforms show the signal which would be observed on an oscilloscope connected to input and output.

It is much easier to relate these signals and their polarities if the positive line is at the top of the drawing, as shown, than if it were drawn below the earth line.

On many maintenance manuals for apparatus concerned with pulse waveforms, for example television receivers and oscilloscopes, the wave forms to be expected at key points in the circuit are shown by the inset drawings. These are always drawn with a positive upwards convention to agree with the manner in which oscilloscopes deflect.

(3) Use standard layouts for particular groups of components which are frequently used together.

7.9   STANDARD COMPONENT LAYOUTS

These groups of components often used together may be regarded as building bricks from which larger circuit diagrams can be assembled. Examples are

> bridge rectifiers
> amplifiers
> oscillators
> multivibrators, etc.

A comprehensive list of these is given in BS 3939: Guiding principles, and a few examples are shown in figures 7.12 and 7.13

The first item of figure 7.12 shows the standard layout for a bridge rectifier, with the input (a.c.) on the left and the output (d.c.) on the right.

Item 2 shows a similar circuit with three-phase input. However, many textbooks and manufacturers' catalogues show drawings of this circuit with the lines to the diode elements vertical rather than at 45°. This makes the circuit easier to draw but not easier to follow.

The diagrams for single-stage amplifiers are drawn with the base biasing resistors on the left of the transistor in items 3 and 4 as the signal input is taken to the base. The output electrode (collector or

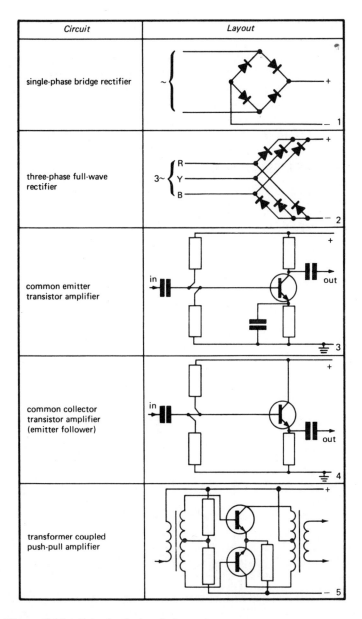

Figure 7.12 Standard circuit layouts

emitter) is drawn on the right-hand side of the transistor symbol, to agree with the convention 'input on the left, output on the right'.

The major feature of push-pull circuits is that they should be drawn in a symmetrical fashion to represent the symmetrical connections of the circuit. Thus in item 5 the top and bottom halves of the diagram are mirror-images of one another, apart from the base bias resistors and the common emitter resistor. Transistor push–pull amplifiers are frequently directly coupled, so avoiding the need for transformers as used here, but the same symmetrical arrangement of the diagram is still used.

Figure 7.13 shows the recommended layouts for a Hartley

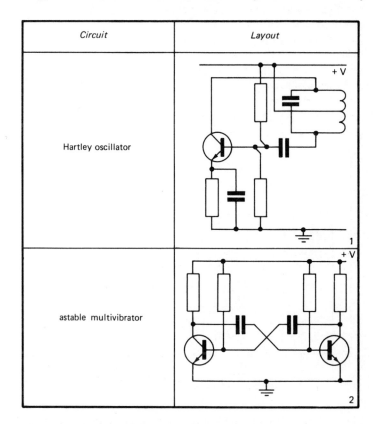

Figure 7.13 Standard circuit layouts

oscillator and an astable multivibrator. The oscillator required a centre-tapped coil in its earliest version, but for a transistor circuit the tapping is usually much closer to the end which feeds the base. It is, however, still the custom to draw the coil, as here, as if it were tapped in the centre.

The astable multivibrator comprises two transistors with identical components cross-connected. The circuit is symmetrical and is drawn as shown so that the two halves are mirror-images of one another.

The same basic layout should be used for the bistable multivibrator, which is completely symmetrical but the transistors are d.c. coupled, not a.c. coupled as in the astable circuit. The monostable multivibrator uses one d.c. and one a.c. as coupling and thus is not entirely symmetrical.

7.10 EXAMPLES OF CIRCUIT DIAGRAMS

Figure 7.14 shows a three-stage directly coupled amplifier which forms the first section of the recording chain for a tape-recorder. Since there are no inter-stage capacitors the stages are connected directly from collector to base and to simplify the layout each transistor is drawn higher on the page than the previous one. As there is a single amplifying signal path the input is on the left and the output on the right.

Figure 7.15 shows the X-deflection amplifier of an oscilloscope

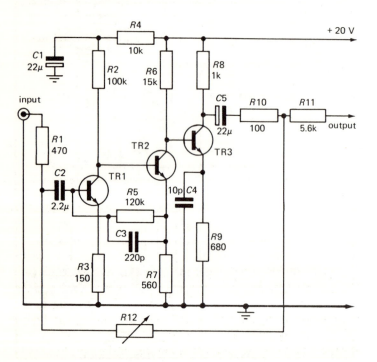

Figure 7.14   Circuit diagram of a MIC/line amplifier for a tape-recorder

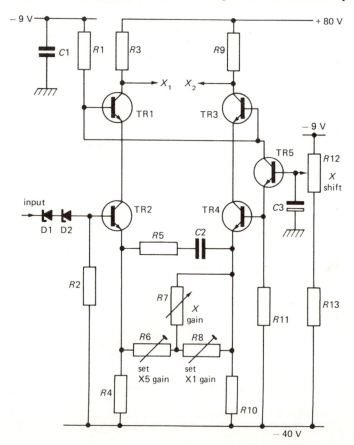

Figure 7.15 X-deflection amplifier for an oscilloscope

used for servicing and fault-finding. The signal path here is somewhat more complicated as the single sawtooth input wave form must generate two push–pull output signals to drive the two X-deflection plates. The circuit has a degree of symmetry and comprises two transistors in each half of a 'long-tailed pair'.

The two emitter electrodes of TR2 and TR4 are coupled via R6 (preset) and R7 (the front panel 'X-gain' control).

The two circuits R3, TR1, TR2, R4 and R9, TR3, TR4, R10 are identical and therefore they are drawn side by the side between the +80 V and −40 V lines.

In this drawing the convention of positive supply line at the top and negative supply line at the bottom is followed with regard to the main +80 V and −40 V lines. The relative positions of the chassis and −9 V lines have however been altered to simplify the layout.

In contrast to the electronic circuits shown in figure 7.14 and 7.15, a power circuit diagram is given in figure 7.16.

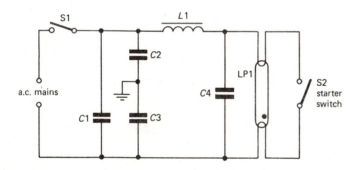

Figure 7.16   Circuit of a fluorescent lamp

This is for a low-pressure fluorescent lamp. Whereas a filament lamp circuit would include only an on–off switch and a lamp, a fluorescent tube requires a power factor correction capacitor C1, a ballast choke L1, R.F. interference suppression capacitors C2, C3 and C4, and a starter switch S2.

The layout is arranged with the a.c. mains input on the left to agree with the normal convention.

7.11 PRODUCING DIAGRAMS FROM EQUIPMENT LAYOUTS

Although circuit diagrams should be provided for all equipment which needs servicing or adjustment, on occasions work must be carried out without them. Unless the task is fairly trivial it is necessary to produce a circuit diagram by tracing out the wiring of the apparatus. As an example of this process we will take the printed circuit board shown in figure 7.17.

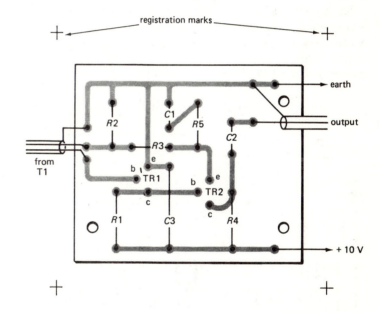

Figure 7.17   Printed circuit board for a transistor amplifier

This is a composite drawing in that the component identities are usually printed on the top or component side of the board, and the copper connection strips are on the reverse side. The view of the board shown in figure 7.17 is that seen by holding the board against a strong light. The insulating board is translucent and the copper strips on the reverse side are seen as dark portions on a lighter background.

The first task is to identify the power supply points by tracing the external connections or by voltage measurement. The transistors can then be identified as *n–p–n* types because their emitters are connected directly or via resistors to the negative line, and their

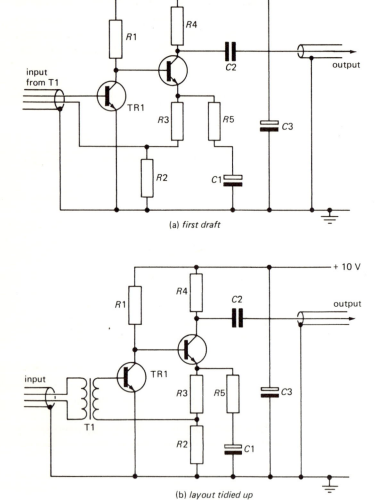

(a) first draft

(b) layout tidied up

Figure 7.18   Circuit diagram drawn from printed circuit of figure
7.17

collectors to the positive line. Alternatively if their type numbers are given, this information can be found from manufacturers' catalogues.

A first sketch of the circuit diagram can then be made by drawing the transistors and then following the connections through from each terminal of the transistors to the earth or + 10 V line.

The result is shown in figure 7.18a. The arrangement of the windings of T1 can also be discovered by testing with an ohmmeter, looking for a diagram on its case, or consulting the makers' catalogue. Having established the connections between components, a second version of the drawing can be made as in figure 7.18b in which the layout is improved and the windings of T1 are shown.

Note that in figure 7.18a the layout of figure 7.17 is turned upside down to conform to the convention 'positive uppermost'. Thus we draw the positive line at the top of the diagram and the earth at the bottom. This is the opposite of the layout in figure 7.17 which has the earth conductor at the top.

In order to make the layout easy to follow the components in figure 7.17 have not been packed together as closely as they could be. For example, the distance between the centres of the pads to which the resistors are connected is at least 11 mm. This could be decreased to 9 mm if 1/4 watt resistors were used, so allowing the dimensions of figure 7.17 to be reduced by about 18 per cent.

Part of a slightly more complicated printed board is shown in figure 7.19. The whole board measures 140 mm × 110 mm and forms part of an obsolescent digital computer. It is a general-purpose logic board that contains six of the NAND circuits shown which have differing numbers of input diodes. The circuit has provision for extending the number of inputs by connecting some extra diodes on the board to the common of D1, D2, D3, and the collectors of the transistors are left open so that two or more can be connected to the same load resistor. Normally each circuit will be connected to its own load resistor, three of which are shown as R4, R5 and R6. For this circuit R4 could be used as the collector load by connecting together pin A and pin 4.

This circuit differs from the one in figure 7.17 because it uses a 'double-sided' board, that is, one which has copper connecting

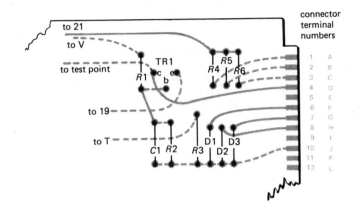

Figure 7.19   Part of a printed circuit board for a logic unit

strips on both sides of the board. In figure 7.19 the strips on the front or component side of the board are shown in full and those on the reverse side are shown dashed.

Although double-sided boards are more complex to make, they give much greater flexibility to the designer by allowing many more 'cross-over points', where one conductor can cross another without being connected to it. In a single-sided board these cross-overs can be provided only by inserting straps of insulated wire between two sections of conductor.

A double-sided board usually has also a double-sided edge connector which plugs into a mating connector to extend the board wiring to the rest of the equipment.

In figure 7.19 the pins of the connector and the corresponding pads of the edge connector are numbered on the top face, and labelled alphabetically on the lower face. Thus the first front pad is labelled 1, and the first lower pad is labelled A. To allow the maintenance engineer to check that the circuit is operating correctly, six test sockets that are accessible from the front of the equipment are fitted to the front edge of the board. Each socket is connected to the collector terminal of one of the six circuits; in this circuit it is pin 4 of the edge connector.

When drawing the circuit diagram the first stage is to copy the layout of the board, but inserting the appropriate symbols for the

components. This gives a diagram similar to that of figure 7.20a.

The transistor type indicates that it is a *p–n–p* device, and the diode polarity can be seen by inspection (a ring is marked on the body at the cathode or lower end in figure 7.20a). This information reveals that the circuit operates as a NAND gate, with negative logic, that is, logic 0 is a signal near earth, and logic 1 a signal near

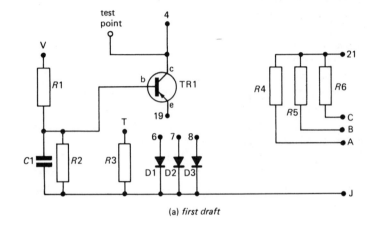

(a) *first draft*

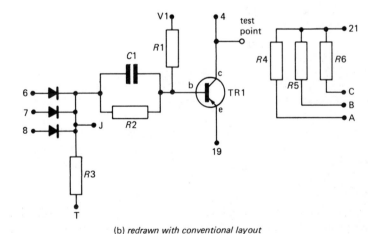

(b) *redrawn with conventional layout*

Figure 7.20   Circuit diagram of a logic unit drawn from figure 7.19

E.D.—E

the voltage of the collector supply on pin 21, −6 V. The circuit can now be redrawn using a conventional layout as shown in figure 7.20b.

The transistor emitter (pin 19) is earthed, pin V is connected to a negative supply, and pin T to a positive supply.

The circuits of most electronic apparatus can be traced in a similar manner, but current digital circuits are more difficult to follow as they consist mainly of integrated circuits. These have pins separated by only 0.1 in. (2.54 mm) and in order to pack in enough conductors the printed circuit tracks may be only 0.8 mm wide, spaced by 0.5 mm.

This spacing is much closer than that shown in figures 7.17 and 7.19 and a magnifying viewer (as used for inspecting the boards after assembly) will help in tracing the circuit.

The printed circuit boards currently used in large computers cannot however be analysed in the same way as they consist of a sandwich of insulating boards and up to six layers of conductor. Visual inspection will not reveal the way in which the buried conductors are connected together, and it is necessary to check each interconnection electrically.

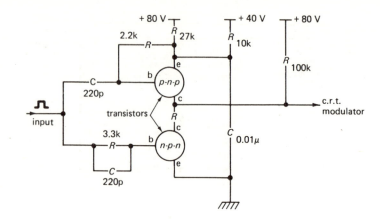

Figure 7.21

EXERCISES

7.1 Figure 7.21 shows the circuit of the pulse amplifier which provides beam suppression during the fly-back interval for a cathode-ray oscilloscope. Draw the circuit using standard BS 3939 symbols for the components.

7.2 Figure 7.22 shows the circuit of the oscillator stage of an F.M. receiver, which is tuned by varying the reverse voltage across the varactor diode. Draw the circuit using standard symbols.

7.3 Figure 7.23 shows the circuit of a mains driven fullwave rectifier and voltage regulator. The voltage reference is obtained from the two Zener diodes ZD1 and ZD2. Draw the circuit using standard symbols.

7.4 Figure 7.24 shows part of the printed circuit card for the

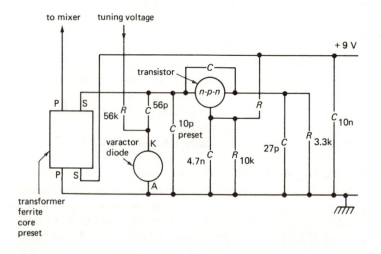

Figure 7.22

remote control amplifier of a television receiver. P1 is connected to the microphone input, and the collector of TR27 feeds at the terminal X a circuit tuned to the ultrasonic frequency used. Draw

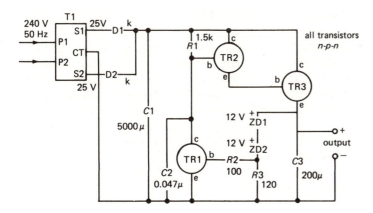

Figure 7.23

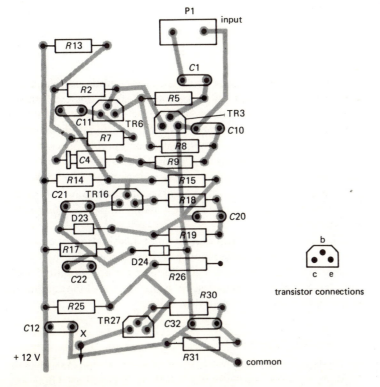

Figure 7.24

transistor connections

a circuit diagram of the four stages shown in figure 7.24.

7.5 The circuit diagram of a d.c.-to-d.c. forward converter is shown in figure 7.25. The transformer T1 has a single primary winding P, and two separate secondary windings S1 and S2. The start and finish of each winding is indicated by the letters S and F. Thus S1S is the start of secondary winding S1. Draw the diagram using standard BS 3939 symbols.

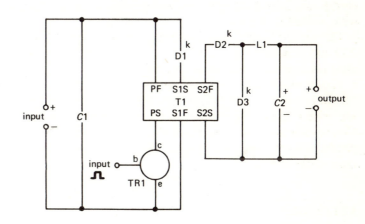

Figure 7.25

8 Equipment Design

8.1 DESIGN ASSESSMENT

A situation which arises frequently in engineering organisations is the selection of one from a number of alternative designs or products. The user provides a list of requirements and technical staff are required to decide which is the most suitable.

Ideally it should be possible to assess the various designs in numerical terms. This requires that the user should attach a weighting factor, say, W_1, W_2, W_3, etc., to each feature which he requires. The assessment then involves assigning a score, say, S_1, S_2, S_3, etc., to each feature of all the designs. The total score for a particular design is then given as the sum of the products

$$S_A = W_1 S_1 + W_2 S_2 + W_3 S_3 +, \text{etc.}$$

The design with the highest score is then the one which best meets the user's requirements.

This kind of assessment is possible in only a few situations where the user is sufficiently certain of his needs to provide firm weighting factors.

Many users are uncertain about their exact needs, particularly in the future if the equipment has a long life, and this exact rating method cannot be used. However, some factors may be regarded as essential, so that the assessment is simply a yes/no decision. If the design has the required feature it may be acceptable, depending on other factors. If it does not, it cannot be suitable and must be eliminated.

For example, any measuring equipment required for use in the field must be battery operated since a mains supply will not usually be available.

Likewise, any equipment used in inflammable atmospheres such as might be found in an oil refinery or down a coal-mine must be of an approved flame-proof design and meet the appropriate British Standards. Unless this requirement is met the use of the equipment is not only risky, but illegal also.

8.2 DESIGN CRITERIA

The features which are important in a particular product depend partially upon the purpose for which it is required. There are, however, a number of factors which are important in all situations.

(1) Technical Performance. Most engineering products have to reach a specified level of performance, often laid down by a British Standard specification. In some applications this is merely a minimum requirement and a product may have added value if its performance exceeds the minimum required.

(2) Reliability. This is the probability that the equipment will operate to its required standard without failure for a specified period. The reliability may be set by purely financial considerations, as for example in the design of television receivers for rental. If these have insufficient reliability the rental company will lose money through the expense of frequent repairs. However, increasing the reliability of a product costs money and if the manufacturer tries to increase the reliability too much the cost of the receiver will become excessive and cannot be recovered from the rental charge.

Some apparatus, such as aviation, electrical and electronic equipment and defence equipment, has to meet the reliability figures specified by the user, and in some contracts any excess reliability shown by testing can lead to bonus payments to the manufacturers. In a few cases the reliability may be specified by legislation or some licensing authority, as is the case for blind-landing equipment for civil aircraft.

(3) Cost. This is always important, but less so in equipment of a complex and critical nature—such as military and aviation equipment—than in domestic equipment where competition is generally much keener.

(4) Ease of Servicing and Adjustment. An increasing proportion of consumer equipment is made as 'throw-away' items which cost more to repair than replace. However, there is considerable resistance to this idea as many products contain parts which will wear out and need replacing several times during the life of the product.

It is important to ensure that these consumable parts are cheap, readily available and can be easily fitted without expensive dismantling of the product. This is particularly important for equipment which is expected to last for many years, and many large authorities now assess equipment on a 'total life cost' basis.

This means that the important feature is not the purchase price but the total cost of ownership. This includes the cost of running the equipment during its lifespan, and paying for spare parts, repairs and routine maintenance.

This concept of total cost can give a completely different assessment from that based on just the purchase price. In the early days of the last war, for example, the American armed forces found that the cost of spares and servicing for some electronic equipment added up during its life duration to over ten times the original purchase price. When assessing such equipment one could almost ignore the purchase price and examine only the subsequent cost of repairs and replacements.

The increasing use of total life cost for evaluating equipment has encouraged the design of equipment which needs the minimum of maintenance and periodic adjustment. In mechanical items examples of this are the use of bearings impregnated with lubricant which will last for the life of the equipment without requiring attention; and in electronic circuits the increasing use of digital methods which require no adjustments is in contrast to linear circuits which usually require preset resistors for setting zero levels, gains, etc.

(5) Size and Weight. The importance of these features depends almost entirely upon the application. For example, any measuring or monitoring equipment used in a power-station will not be restricted unduly since the size and weight of the equipment is negligible compared with that of the generators in the station. However, both items are most important in airborne equipment where each pound which the equipment weighs represents a pound less of payload and hence of earning capacity. Size is also important because a great deal of apparatus must be packed into a very limited space.

Weight is extremely important in any product which is supposed to be portable—particularly if this is for the consumer market. So called 'portable' radio receivers were made in the 1920s but they were clumsy because they used thermionic valves and so required heavy batteries. Consequently, relatively few were sold.

When cheap transistors became available, small and much lighter receivers could be made and these became so popular that they captured a large proportion of the total receiver market.

(6) Power Consumption. Again this feature may be comparatively unimportant in places such as power-stations or steelworks where large amounts of power are handled. It becomes more important in airborne equipment and in apparatus such as remote or submerged telephone repeaters where power may have to be supplied from a source many hundreds of miles away (for a transatlantic cable for example), or powered from solar cells. Probably the most critical application is in satellite equipment where power supplies are very limited and low power consumption is required at the expense of other features.

8.3 COMPONENT SELECTION

An important factor in the design of equipment is the correct selection of the appropriate type of component for each position in the circuit. As with the design of the complete equipment, a suitable compromise between cost and performance is required and an important factor is the assessment of the quality and stability required for each item. These may vary very considerably between different circuit functions. For example, a resistor required for decoupling the power supply to a transistor could be allowed a variation of at least ± 10 per cent without causing any loss of performance. At the other extreme, a resistor in a precision D–A converter may need a stability of perhaps ± 0.05 per cent despite temperature changes, and most introduce very little noise into the circuit. In the following sections we consider the various types of resistor and capacitor and the circumstances in which they are generally used.

8.4 TYPES OF RESISTOR

The fixed resistor is the most widely used component in the majority of electronic equipment. The possible exception is digital apparatus in which integrated circuits are used. These can be connected directly to one another and so require no resistors for coupling or biasing. The main features of a resistor that are important to the designer are cost, power dissipation, stability and noise level.

The cheapest resistor is the carbon composition type. It is, however, liable to drift in value during its life, particularly those over $1 M\Omega$, and if passing a direct current it introduces extra noise into the circuit. For these reasons it is used only for non-critical situations in consumer equipment.

A more stable and less noisy device is the carbon film resistor (sometimes called a 'high-stability' resistor) and this is the general purpose resistor for industrial equipment. It consists of a thin layer of carbon on a ceramic rod former. It must be hermetically sealed if used in a humid atmosphere as moisture can cause rapid corrosion.

More accurate and stable resistors are now available using a metal or metal oxide coating on a glass or ceramic substrate, but they are somewhat more expensive than carbon film resistors.

Finally, for extremely accurate measurements precision wirewound resistors are used. They can be obtained with tolerances of ± 0.05 per cent or ± 0.1 per cent, and with temperature coefficients of around 5 parts per million per °C. The cost is about a hundred times that of a carbon composition resistor with a tolerance of ± 10 per cent when supplied.

All of the above types are low-power devices, which can dissipate up to $\frac{1}{2}$ W generally, and in some cases up to 2 W. For greater dissipation (up to about 20 W) wirewound resistors with a ceramic former, protected by a coat of vitreous enamel, are used.

For still greater power dissipation, for example in starters for electric motors, wirewound resistors on formers of vitreous-enamelled steel tube are often used. Other constructions for high-power resistors include cast grids and coiled iron wire. These may be cooled naturally by convection or by forced air cooling.

8.5 TYPES OF CAPACITOR

A very wide range of capacitors are available for various purposes, the cheapest per micro farad being the electrolytic type. This is restricted to use for smoothing and decoupling power supplies, and in the smaller sizes for interstage coupling. The important requirement is that it must always have one plate more positive than the other or else its capacitance changes drastically. Other limitations are that it always has a small temperature-dependent leakage current, and its tolerances on capacitance is typically − 25 per cent + 100 per cent.

Where used as a reservoir or smoothing capacitor in power units incorporating mains energised rectifiers it may have to carry high ripple currents, up to 10 A or more, and special types are made for this duty. Very compact low-leakage capacitors can be made using Tantalum rather than aluminium electrodes, but these are limited to a working voltage of about 40 V, whereas the normal aluminium types have working voltages up to 450 V. A further limitation is that electrolytic capacitors are designed to work up to temperatures of only about 70 °C, whereas military equipment may need to operate at temperatures up to 125 °C.

Because of their construction electrolytic capacitors have some inductance and they will thus not act as bypass capacitors very effectively at high frequencies.

For low-leakage application in values up to 10 μF various plastic films are used to make capacitors (for example polycarbonate and polyester). These have low dielectric losses and stable capacitance values and are thus suitable for use in tuned circuits. For small values, up to about 0.01 μF, capacitors made by depositing silver films on to a thin sheet of mica are still used as they have low temperature coefficients. They are, however, several times more expensive than polystyrene film capacitors which are electrically similar.

Finally for decoupling at very high frequencies, for instance above 20–50 MHz, capacitors using a high permittivity ceramic dielectric are used. These are very small, of plate construction and are thus of very low inductance. Different types of ceramic are used to give high dielectric strength, so enabling very compact capacitors to be made which will withstand high voltages of the order of 10 kV.

8.6 INSULATING MATERIALS

It is difficult to imagine any electrical or electronic product which does not involve the use of some insulating material in its construction. Thus the selection of a suitable insulating material is important if the product is to perform its function satisfactorily.

The two basic factors necessary for satisfactory insulation are as follows.

(1) The material must have a very high resistance so as not to pass any appreciable leakage current.

(2) It must also have a high dielectric strength so that only a thin layer can withstand a high voltage without breakdown. This requirement is most important in all high-voltage equipment, but becomes much less necessary in low-voltage apparatus. For example in 10–20 V transistor circuits, or the switches and wiring of a car, the thickness of the insulation is usually determined by the need for mechanical strength rather than electrical breakdown, and would suffice for a few hundred volts.

One of the problems with the selection of insulating materials is that they normally have to perform tasks other than merely sustaining a difference of potential. They are normally required at least to support conductors, as in overhead power lines or substation busbars. Secondly, the material must meet a mechancial requirement as well as an electrical one. The same holds for the dolly and other moving parts of a switch or contact breaker, or for the insulation of the windings in a motor or generator. Although most of the force in these machines is exerted on the iron of the rotor, there is still a considerable driving force exerted on the conductors as well as the large centrifugal force as a result of high speed rotation.

Furthermore, most insulating materials in power equipment have to conduct heat from the current-carrying conductors, so their thermal conductivity is of interest. Alternatively, the insulating material may be circulated to conduct the heat away bodily, as with oil-cooled power transformers or hydrogen-cooled alternators. Here the material should have a high thermal capacity to transport the heat and a low viscosity so that it needs a minimum of power to drive it around the cooling circuit. Hydrogen is markedly better than air on both these features, which accounts for its use instead of air for cooling large alternators.

8.7 LOW-VOLTAGE INSULATION

For many purposes in low-voltage electrical and electronic equipment the voltage stresses are not a major factor, and the selection of insulating material depends much upon cost and ease of fabrication. Plastics are widely used because intricate shapes—for example the body of a toggle switch or a wafer switch—can be moulded cheaply and at high production speeds. Where the plastic is moulded around conductors, for example in transistors, an important feature is that the insulation should 'wet' the metal and so adhere to it, thus keeping out moisture in humid conditions. Where large temperature variations occur, the coefficient of thermal expansion of the insulation should match that of the metal it is bonded to fairly well, otherwise a change of temperature may separate them and allow contamination.

Where insulation is required in the form of thin sheets, plastic materials alone are generally too weak, and so they are strengthened by using paper, cloth or glass fibre, as for example in printed circuit boards.

Where space is important, as in transformer windings, there is an economic advantage in using materials which have a high dielectric strength, so that the minimum thickness of insulation is needed. Where the temperature rise is not too large, organic materials such as paper and cotton tape may be used but for high-temperature working synthetic materials such as glass-fibre tape are needed.

Also, for insulating the wire used in the windings a plastic film is now generally preferred to the older enamel. It will work at higher temperatures, and a given thickness will withstand a much greater voltage between adjacent turns.

In order to speed up the task of connecting the ends of the windings to tags on the transformer or coil bobbin, 'self-fluxing' coatings are available. When using a soldering iron which is rather hotter than usual, the insulation melts and acts as a flux to assist making the joint.

For high-voltage insulation in colour television receivers and measuring equipment, polythene is widely used as it has a high dielectric strength, resists moisture, and can easily be moulded.

8.8 MATERIALS FOR HIGH TEMPERATURES AND HIGH FREQUENCIES

Some insulating materials, for instance those used in electric heating and in the heater/cathode assembly of thermionic valves and cathode-ray tubes, need to operate at red heat and remain mechanically stable. For this function the materials most used are ceramics and mica—a naturally occurring mineral. For some applications natural mica is too thin, so small particles of mica are compacted with a binder into tubes and sheets. Natural asbestos or asbestos moulded into sheets and tubes is also used.

The electrical conditions we have considered so far have been those where insulation is subjected to d.c. or low frequency a.c. fields. The only factor which causes losses in the dielectric is then leakage. However, higher frequencies are being increasingly used for many purposes and here a further cause of dielectric loss and the resulting internal heating arises.

This is because the energy stored in the dielectric when it is subjected to a voltage stress is not all returned to the supply when the stress is removed. The balance appears as heat in the dielectric. This loss is similar to the loss due to magnetic hysteresis when a ferromagnetic material is magnetised, and like this hysteresis loss its magnitude increases with frequency.

As a result, the selection of insulating materials becomes more critical as the working frequency is increased. The heating effect is of course used deliberately in some applications such as R.F. curing and welding of plastics, and microwave ovens.

Generally, naturally occurring materials are less satisfactory than synthetic ones. Air or high vacuum are satisfactory, as are some materials such as P.T.F.E., but older plastics such as bakelite are useless. Wherever possible the minimum amount of insulation is used (by using thin sheets or foamed material) so that most of the volume in which the electric field exists is occupied by air.

8.9 EQUIPMENT TESTING

An important feature of much design assessment is the testing of new apparatus to ensure that it meets the user's requirements. This may involve an extremely wide range of activity. At one extreme, for example with a new tool for stripping the insulation from plastic-coated wires, simple tests will suffice. Thus one could compare the time taken to use the tool with that required by previous methods, check that it caused no damage to the wire itself, and then hand it over to a technician for a few weeks to obtain a view of its convenience and speed under working conditions.

On the other hand, equipment intended for aerospace use may require extensive environmental testing, such as being subjected to alternate cycles of heating and cooling, periods of exposure to humid heat and radiant heat, reduced pressure to simulate high altitude operation, etc. The electrical performance must be monitored during these tests. In addition the equipment must be subjected to controlled vibration at a range of frequencies.

These tests ensure that the apparatus will operate in all environments to which it is likely to be subjected. Further long duration tests are also carried out on production models to assess their reliability.

Clearly, facilities for this kind of testing are very expensive and only large organisations can afford them. However, the testing of small items for the consumer market requires much simpler facilities since the environment is much less severe and the reliability needed is considerably less. Electronic units can be put under full-load conditions, periodically switched on and off, and their performance checked at regular intervals. Electromechanical apparatus such as switches, relays, contactors, etc., are put under electrical load and operated repeatedly for 10^5 or 10^6 operations, the contact resistance being monitored periodically. In some cases it may be useful to continue the test until failure.

A particular difficulty with many modern components is their extreme reliability and the need for accumulating millions of component-hours of testing to obtain useful reliability data under moderate working conditions. There is thus a place for tests in conditions which are much more severe than those ever encountered in service, so that failures will occur much more quickly. Although it is not always possible to relate tests like these ('accelerated' life tests) to the probable behaviour under working conditions with much accuracy, they are very valuable in production testing in order to compare quickly the quality of one

batch of components with that of previous batches.

8.10 ASSURED QUALITY COMPONENTS

A considerable volume of electrical and electronic equipment is now being used in military and aerospace systems where reliability is very important. To avoid every user having to produce his own (different) specifications for component quality, a national scheme has been started which lays down several different levels of environmental conditions and the corresponding methods of testing. Certain manufacturers and suppliers who can demonstrate that they have adequate facilities for quality assurance and testing have qualified for producing certain types of assessed quality components. The scheme is often called the BS 9000 scheme since all of the specifications and standards are numbered from 9000 onwards.

A similar scheme to cover the European market is now being developed by the international standards body CENELEC. The result is that users can specify components to the appropriate standard and have the benefit of a steadily growing volume of testing of the components under known conditions. This provides an increasingly accurate knowledge of the failure rates likely to be obtained in working conditions.

QUESTIONS

8.1 When selecting equipment for a particular purpose the items which do not satisfy the electrical requirements are eliminated. When choosing from the remainder the size, weight, cost, reliability and power consumption are assessed. Put these features in order of importance for

    (a) the mains energised battery charger for a calculator
    (b) a public address amplifier for a civil airliner
    (c) a starter for a 50 hp motor to be used in a steelworks.

What other features do you consider important in each case?

8.2 You are asked to set up a life-test rig for the mains switch used in a domestic radio receiver. How would you estimate what electrical conditions should be imposed on the switch, and how many operations would you propose for the test?

8.3 For many years general measurements of voltage, current and resistance in service workshops have been made using analogue multimeters with a pointer and scale based upon a moving-coil movement. Recently digital multimeters have been introduced which give a reading in digital form using light-emitting diode or liquid crystal displays.

Draw up a list of the features you consider important in selecting a multimeter for a small electronic service workshop and compare the two types of meter. Which would you prefer to use and why?

Appendix A — Standards in Electrical Drawing

1. NATIONAL STANDARDS

The most important British Standards concerned with drawings of electrical and electronic equipment are:

BS 308:1972: Engineering drawing practice
BS 3939: Graphical symbols for electrical power, telecommunications and electronics diagrams
BS 5070:1974: Drawing practice for engineering diagrams
I.E.E.: Wiring regulations, 14th edition

The first of these, BS 308, deals only with mechanical drawings but its recommendations about drawing sizes, title blocks, numbering and lettering are usually adopted for electrical diagrams also.

BS 3939 has been issued in 29 parts at various times between 1966 and 1979. It deals with symbols for electrical and electronic components and assemblies and their use in various kinds of diagram, including circuit diagrams and block diagrams.

BS 5070:1974 includes some electrical and electronic diagrams and also diagrams of pipe work, hydraulic, and other engineering systems. It covers diagrams for mechanical, civil and chemical engineering in addition to those for electrical engineering.

Appendix B contains a list of other British Standards which deal with components and materials that are used in electrical and electronic equipment.

When preparing any drawings involving electrical installations the I.E.E. Wiring Regulations should be consulted. Although they include few drawings, they specify in much detail the size and capacities of cable ducts, the sizes and ratings of electrical cables and conductors, the ratings of fuse wires, the colour code for various circuits, etc. They are revised periodically, the present version dating from 1974.

2. INTERNATIONAL STANDARDS

Although standards were originally proposed by national bodies, international trading soon showed the importance of international standardisation. This is now in the hands of the

International Standards Organization (I.S.O.), which delegates responsibility for standards in certain fields to smaller bodies. The major organisation for electrical and electronic standards is the International Electro-technical Commission (I.E.C.).

In addition to recommending standards in fields not previously covered, the I.E.C. has also been working towards a common international agreement in cases where national standards originally differed.

An example of this 'harmonisation' process occurred with many of the symbols for circuit diagrams which are included in BS 3939. Some years ago there were at least three widely used sets of symbols, which had been developed in the United States, the United Kingdom and the Federal German Republic.

In the intervening years the I.E.C. has persuaded these and other countries to adopt a common set of symbols which are now generally used. It is still necessary to recognise the earlier symbols, however, as many equipment diagrams which embody them have not been revised.

The present version of BS 3939 includes a note alongside each symbol which shows whether it is also an I.E.C. standard symbol. Most of them now come into this category, although a number deal with components for which the I.E.C. have yet to issue recommendations.

As an example of the way in which BS 3939 symbols have changed over the years, the symbol for logic gates is interesting. This was originally a circle, being defined as a purely British standard. As a result of discussions with other European countries it was changed to a semicircle, and then the I.E.C. finally decided that a rectangle would be more suitable, mainly because it is easier to draw. This is the shape which now appears in the latest version of BS 3939, section 21, and in the logic diagrams of BS 5070.

Appendix B — British and International Standards

1. BRITISH STANDARDS CONCERNED WITH ELECTRONIC AND ELECTRICAL ITEMS

Number	Title
31:1940	Steel conduit and fittings for electrical wiring.
52:1963	Bayonet lamp-caps, lamp-holders and B.C. adaptors (lamp-holder plugs) for voltages not exceeding 250 V.
88:1975	Parts 1 and 2. Cartridge fuses for voltages up to and including 1 kV a.c. and 1.5 kV d.c.
91:1973	Electric cable soldering sockets.
158:1961	Marking and arrangement of switchgear bus-bars, main connections and small wiring.
159:1957	Busbars and busbar connections.
229:1957	Flame-proof enclosure of electrical apparatus.
448:	Dimensions of electronic tubes and valves.
546:	2-pole and earthing pin plugs, socket outlets and socket outlet adaptors for up to 250 V.
741:1959	Flame-proof electric motors.
1361:1971	Cartridge fuses for a.c. circuits in domestic and similar premises (similar to I.E.C. 269–1).
1363:1967	13 A plugs, switched and unswitched socket-outlets and boxes.
1454:1969	Consumers electricity control units (up to 100 A).
1568	Magnetic tape-recording equipment. Part 1:1970. Magnetic tape-recording and reproducing systems, dimensions and characteristics (equivalent to I.E.C. 94). Part 2:1973. Cassettes for commercial tape-recorders and domestic use, dimensions and characteristics (equivalent to I.E.C. 944).
1852:1975	Marking codes for resistors and capacitors (equivalent to I.E.C. 62).
1927:1953	Dimensions of circular cone diaphragm loud-speakers.
1991:	Letter symbols, signs and abbreviations. Part 6. Electrical science and engineering (equivalent to

	I.E.C. 27–1). Supp. 1, Part 6. List of subscripts for electrical technology (equivalent to I.E.C. 207–1 and 27–2).
2048:	Dimensions of fractional horsepower motors. Part 1:1961: Motors for general use.
2131:1965	Fixed capacitors for direct current using impregnated paper on paper/plastics film dielectric.
2136:1965	Fixed metallised-paper dielectric capacitors for direct current.
2251:	Sockets for electronic tubes and valves.
2258:	Shields for electronic tubes and valves.
2316:	Radio frequency cables.
2401:1966	Polyester film dielectric capacitors for direct current for use in telecommunication and allied electronic equipment (similar to I.E.C. 202).
2470:1973	Hexagonal socket screws and wrench keys (inch series).
2517:1954	Definitions for use in mechanical engineering.
2754:1976	Memorandum. Construction of electrical equipment for protection against electric shock.
2757:1956	Classification of insulating materials for electrical machinery and apparatus on the basis of thermal stability in service.
2771:1974	Electronic equipment of machine tools. General-purpose and mass-production machines and their electronic equipment (similar to I.E.C. 204–1, 204–2 and 204–3).
2856:1973	Precise conversion of inch and metric sizes on engineering drawings (similar to I.S.O. 370).
2917:1977	Specification for graphical symbols used on diagrams for fluid power systems and components (equivalent to I.S.O. 1219).
2950:1958	Cartridge fuse-links for telecommunications and light electrical apparatus.
3041:	Radio-frequency connectors, Part 2:1977 specifications for coaxial unmatched connector (equivalent to I.E.C. 169–2).
3238:	Graphical symbols for components of servo-mechanisms.
3283:	Non-reversible connectors and appliance inlets for portable electric appliances (for circuits up to 250 V).
3363:1968	Schedule of letter symbols for semiconductor devices (similar to I.E.C. 148).
3429:1975	Sizes of drawing sheets.
3456:	Specification for the safety of household electrical appliances (42 sections covering a wide range of products).
3643:	I.S.O. Metric screw threads.
3676:1973	Switches for domestic and similar purposes (for fixed or portable mounting).
3685:1963	Enamelled and silk-covered copper conductors.
3692:1697	I.S.O. Metric precision hexagon bolts, screws and nuts.
3861:	Electrical safety of office machines.
3901:	Rayon-covered copper conductors.
3902:	Enamelled and rayon-covered copper conductors.
3934:1965	Dimensions of semiconductor devices.
3955:	Electrical controls for domestic appliances.
3979:1966	Dimensions of electric motors (metric series).
4025:1966	The general requirements and methods of test for printed circuits.
4145:1967	Glass mica boards for electrical purposes.
4200:	(8 Parts) Guide on the reliability of electronic equipment and parts used therein.
4293:1968	Current-operated earth-leakage circuit breakers.
4343:1968	Industrial plugs, socket-outlets and couples for a.c. and d.c. supplies.
4516:	Enamelled copper conductors (P.V.A. base with high mechanical properties).
4520:	Enamelled copper conductors (polyurethane base with solderable properties).
4568:	Steel conduit and fittings with metric threads of I.S.O. form for electrical installations.
4584:	Metal-clad base materials for printed circuits.
4607:	Non-metallic conduits and fittings for electrical installations.

4727:	Glossary of electrotechnical, power, telecommunications electronics, lighting and colour terms.
4808:	L.F. cables and wires with P.V.C. insulation and P.V.C. sheath for telecommunication.
4999:	General requirements for rotating electrical machines.
5000:	Rotating electrical machines of particular types or for particular applications.
5054:1794	Dimensions of spindle ends for manually operated electronic components (similar to I.E.C. 390).
5057:1973	Snap-on connectors.
5069:	Dimensions of piezo-electric devices.
5370:1976	Guide to printed wiring (design, manufacture and repair).
6004:1975	P.V.C. insulated cables (non-armoured) for electric power and lighting.
6007:1975	Rubber-insulated cables for electric power and lighting.
6207:	Mineral-insulated cables.
6231:	P.V.C.-insulated cables for switchgear and control gear wiring (similar to I.E.C. 228).
6346:1969	P.V.C.-insulated cables for electricity supply.
6360:1969	Copper conductors in insulated cables and cords.
6480:	Impregnated paper-insulated cables for electricity supply.
6500:1975	Insulated flexible cords (similar to I.E.C. 227, I.E.C. 245).
6746:1976	P.V.C.-insulation and sheath of electrical cables (similar to I.E.C. 540).
6791:1969	Aluminium conductors in insulated cables.
9000:	General requirements for electronic components of assessed quality.
9001–9760	Deal with particular types of assessed components and methods of testing.

2. INTERNATIONAL STANDARDS FOR TECHNICAL DRAWINGS AND COMPONENTS

I.S.O./R 128–1959	Engineering drawing—principles of presentation.
I.S.O./R 129–1959	Engineering drawing—dimensioning.
I.S.O./R 406–1964	Inscription of linear and angular tolerances.
I.E.C. 27	Letter symbols to be used in electrical technology.
I.E.C. 117	Recommended graphical symbols.
I.E.C. 63	Preferred number series for resistors and capacitors.
I.E.C. 127	Cartridge fuse links for miniature fuses.
I.E.C. 257	Fuse-holders for minature cartridge fuse links.
I.E.C. 62	Colour code for fixed resistors.

Bibliography

Abbott, W., *Technical Drawing*, 4th edition (revised by T. H. Hewitt) (Blackie, Glasgow, 1976).

A.S.E.E., *Illustrated Guide to the I.E.E. Regulations for the Electrical Equipment of Buildings* (A.S.E.E., Leatherhead, 1976).

Astley, P., *Engineering Drawing and Design II* (Macmillan, London and Basingstoke, 1978).

Bishop, G. D., *Electronics II* (Macmillan, London and Basingstoke, 1977).

Heard, W. E., *Intermediate Engineering Drawing* (Macmillan, London and Basingstoke, 1975).

Hewitt, D. E., *Engineering Drawing and Design for Mechanical Technicians* (Macmillan, London and Basingstoke, 1975).

Lopez, U. M., and Warrin, E. M., *Electronic Drawing and Technology* (Wiley, New York, 1978).

Neidle, M., *Electrical Installations and Regulations* (Macmillan, London and Basingstoke, 1974).

Neidle, M., *Basic Electrical Installations*, 2nd edition (Macmillan, London and Basingstoke, 1979).

Richter, H. W., *Electrical and Electronic Drafting* (Wiley, New York, 1977).

Settlement Planning and Participation under Principles of Pluralism

TOVI FENSTER

*Department of Geography, Ben Gurion University, P.O. Box 653,
Beer Sheva 84-105, Israel*

PERGAMON PRESS

OXFORD · NEW YORK · SEOUL · TOKYO

0305-9006(1993)39:3;1-0

Biography

Dr Tovi Fenster is currently Post-Doctoral Fellow in the Ben Gurion University of the Negev, Israel and also professional consultant to Government Ministries and various Regional Councils in the field of social planning. Previous positions held include working for ten years as a social and regional planner in development projects in Israel (mainly with the Bedouin) and in Nigeria, Mexico, Costa Rica and Nepal. Dr Fenster received her Ph.D. in social planning and geography from the London School of Economics and Political Science in 1992.